Environmental and Material Flow Cost Accounting

ECO-EFFICIENCY IN INDUSTRY AND SCIENCE

VOLUME 25

For other titles published in this series, go to
www.springer.com/series/5887

Environmental and Material Flow Cost Accounting

Principles and Procedures

Christine Jasch

Springer IÖW EMAN

Christine Jasch
Institut Für Ökologische Wirtschaftsforschung
Vienna, Austria

ISBN: 978-1-4020-9027-1 e-ISBN: 978-1-4020-9028-8

Library of Congress Control Number: 2008936133

The EMA Assessment templates in Excel and the Brewery Murau Case Study are available for download at www.ioew.at and www.springer.com/978-1-4020-9027-1.

Printed on acid-free paper

springer.com

Foreword

Pressures and incentives for the adoption of cleaner production or pollution prevention processes by business have emerged from both inside and outside enterprises. Internally, the adoption of cleaner technologies may be driven by efforts to avoid the costs of waste management, to bypass the uncertainty of constantly changing regulations, and to position the firm as a "green" enterprise in the local, national, or global marketplace. Externally, corporate environmental performance is increasingly scrutinized by investors, financial advisors, regulatory bodies, host communities, and the public at-large.

In this context, Environmental Management Accounting (EMA) serves as a tool to realize and understand the full spectrum of the environmental costs of non-prevention approaches and the economic benefits of pollution prevention or cleaner ones and to integrate these costs and benefits into day-to-day business decisions.

Environmental Management Accounting is an essential business tool for creating internal demand in businesses for cleaner and less wasteful production processes. EMA changes the order of the reasons why companies may engage in pollution prevention activities from one of environmental concern or market access to market to one of giving a preferential position to engaging in pollution prevention activities purely because it makes good business sense due to the immediate financial benefits it delivers.

If all companies in a national economy were to realize that producing waste is almost always more costly than treating and disposing of it, then without question, these industries would engage in a process of cost reduction through waste minimization rather than focusing on end-of-pipe solutions. This internal demand for cleaner processes would produce nearly immediate changes in overall national waste and emission levels. This process would additionally move companies to strive for continuous improvements in this area as a way to improve profit and efficiency levels and not only as a way to comply with environmental regulations.

The United Nations Expert Working Group on Improving Government's Role in the Promotion of Environmental Management Accounting was organized as a follow up to informal discussions on the issue at the 1998 session of the United Nations Commission on Sustainable Development (CSD 6). The Expert Working Group met nine times between 1999 and 2005 in eight countries and three continents. The members of the Group consisted of experts from national environment agencies and ministries from over 40 developing and developed countries, international organizations, industry, accounting firms, academia, and United Nations agencies.

The purpose of this Expert Group was to support governments in establishing EMA as a viable option for ensuring that the business sector has reliable accounting procedures to assess the true costs of producing wastes and emission and thus is able to better identify the opportunities to improve the efficiency of materials management within production processes, thus reducing wastes, while at the same time being fully cognizant of the financial benefits that these activities include. Within this role one specific target was to bring rigor to the practice of EMA by offering a set of principles and procedures for EMA based on commonly used and internationally accepted financial accounting methods while establishing the boundaries that bind it as an integral part of the internal management process of a company.

The Group succeeded in establishing a common definition and range for EMA while supporting the development of a large number of EMA promoting activities in many countries (UNDSD, 2001; UNDSD, 2002). However, the Group exceeded expectation when it was asked by the Board of Directors of the International Federation of Accountants (IFAC) to cooperate in the development of the IFAC Guidance Document on EMA which aimed to address the deficiencies that the accounting profession had identified in the prevailing accounting procedures in regards to accounting for environmental costs and the costs of producing wastes and emissions.

This watershed publication (IFAC, 2005), achieved the integration of EMA into day to day accounting procedures and thus moved EMA from the environmental world within which it existed since it's inception to the world of the accounting profession were it rightly belongs and were our aim for broad EMA use by every corporate finance office can indeed be achieved.

Having succeeded in the integration of EMA into accounting practice, the Expert Group ceased to exist in 2006 but the influence of its work has continued to grow. Currently a drafting group of the International Standardisation Organisation (ISO), TC-207, is working to develop a new standard on Materials Flow Cost Accounting within the ISO-14000 series on Environmental Management which derives its basic concepts from the Expert Groups EMA definitions and the IFAC Guidance document and the Statistics Division of the department of Economic and Social Affairs of the UN has begun exploratory work on the influence of EMA on the way national statistic on both industry and environment are collected and how the United Nations Handbook of National Accounting – Integrated Environmental and Economic Accounting (commonly referred to as SEEA), would have to be changed to make accessible the benefits that EMA accounting procedures could bring to national statistical systems.

This book aims to condense the accumulated knowledge from the previous work on EMA and join it with the practical experience in the application of EMA accumulated by Ms. Jasch and many others. This publication will be of invaluable benefit to accountants and financial analysts to increase their own value to their respective organizations by providing a practical and business relevant way to asses the financial looses that can be linked to historically inefficient production processes and the potential financial benefits that preventive, environmentally conscious alternatives may provide.

The value of EMA in establishing a culture of pollution prevention and waste minimization within industry is clear. However, the success of government and corporate programs to promote EMA depends on developing EMA systems that are cost-effective for industry.

Ms. Christine Jasch has been a leader in this process from its inception as well as one of its more productive and innovative practitioners. Considering the accelerating growth of EMA practice worldwide and the expected expansion of its influence well beyond managerial costs accounting, Ms. Jasch is singularly qualified to present this compilation of experience and practice which undoubtedly will be of great value to those hoping to find the truth about the profitability of environmentally conscious production processes.

Tarcisio Alvarez-Rivero
Economic Affairs Officer
United Nations, Division for Sustainable Development

Acknowledgments

I would like to acknowledge gratefully the efforts of Mr. Tarcisio Alvarez-Rivero of the Division for Sustainable Development of the United Nations Department of Economic and Social Affairs (DSD/UNDESA), who initiated the UN/DESA EMA Working Group and the IFAC Guidance document on environmental management accounting (EMA) as well as Mr. Hans Günther Schwarz of the Austrian Ministry of Transport, Innovation and Technology, who funded the work necessary to develop the publications within UN DESA and IFAC, on which this book is based.

I would also like to thank the people and companies who participated in the case studies, applied the methodology and tools and helped me develop them at the same time. I have received great feedback from working with Hans Schnitzer, Technical University Graz, and Deborah Savage, Tellus Institute on developing the conceptual framework as well as from Johann Tanner from Obermurtaler Brauereigenossenschaft, Lars Munkoe and Lilian Harbak from Danisco, Walter Hennerbichler and Rudolf Helm from SCA Laakirchen, Rosa Zehner from OMV, Diana Ditulescu and Remus Laes from Petrom and Otto Simon from Verbundgesellschaft in the actual application. I am also grateful for their approval of having the experiences we gained published. My thanks also go to the people and companies of the projects in the Basque country, Costa Rica and Lithuania, especially to Ander Elgorriaga, Myrtille Danse and Zaneta Stasiskiene.

In a current project with UNIDO the COMFAR tool for investment appraisal was enlarged by an EMA assessment tool, which will assist the Cleaner Production Centers as well as their clients throughout the world in realizing the benefits of combined environmental and material flow cost accounting. I am therefore thankful to Elisa Tonda from UNIDO.

Last but not least I would like to thank my partner, Leopold Bernhard, for his patience and understanding for the time I spent with the book instead of with him.

June 2008

Christine Jasch
Institute for Environmental Management
and Economics
IÖW, Vienna
www.ioew.at

Contents

List of Abbreviations

ABC	Activity Based Costing
CA	Cost Accounting
CC	Cost Center
CEPA	Classification of Environmental Protection Expenditure Activities
CP	Cleaner Production
EA	Environmental Accounting
EBIT	Earnings before Interest and Taxes
ECI	Environmental Condition Indicator
EMA	Environmental Management Accounting
EMS	Environmental Management System
EPI	Environmental Performance Indicator
EST	Environmental Sound Technologies
FA	Financial Accounting
FCA	Full Cost Accounting
FEE	Federation des Experts Comptables
GRI	Global Reporting Initiative
HSSE	Health, Safety, Security and Environment
IFAC	International Federation of Accountants
IIR	Internal Interest Rate
ISO	International Standardization Organization
KPI	Key Performance Indicator
MA	Management Accounting
MFCA	Material Flow Cost Accounting
MPI	Management Performance Indicator
NPO	Non-product Output
NRA	National Resource Accounting

OPI	Operational Performance Indicator
SEEA	System of Environmental and Economic Accounting
SNA	System of National Accounts
UNDSD/UNDESA	United Nations Division on Sustainable Development, Department of Economic and Social Affairs
UNIDO	United Nations Industrial Development Organization

List of Figures

List of Tables

Executive Summary

Environmental protection – along with the related costs, revenues and benefits – is of increasing concern to many countries and organizations around the world. Disclosure of related information is requested from several stakeholders (national statistical agencies, financial, environmental and sustainability reporting, climate change emission reports) and thus the necessity for consistent data with auditable data quality is increasing. But there is a growing consensus that conventional accounting practices simply do not provide adequate information for environmental management purposes.

To fill in the gap, Environmental Management Accounting (EMA) emerged with a focus on harmonizing approaches and definitions and providing guidance for corporate implementation. EMA specifically deals with the information necessities for environmental management approaches that bring about benefits to the companies bottom line as well as for environmental performance by highlighting prevention approaches, integrated cleaner technologies and improvements in material and energy efficiency.

The Expert Working Group on Improving Government's Role in the Promotion of Environmental Management Accounting (EMA) was organized as a follow up to informal discussions on the issue at the 1998 session of the United Nations Commission on Sustainable Development (CSD 6) in the context of negotiations on environmentally sound technologies. The participants in the Expert Working Group came from national environment agencies and ministries, international organizations, industry, accounting firms, academia, and United Nations agencies, as well as from the United Nations Division for Sustainable Development. The publication on "Environmental Management Accounting: Procedures and Principles" (UNDSD 2001), was the first of a series of publications by the Expert Working Group, and presents the terminology and techniques as used by members of the group in order to establish a common understanding of thc basic concepts of EMA and provide a set of principles and procedures to guide those interested in its application.

The International Federation of Accountants (IFAC) in 2003 decided to commission a guidance document to bring together some of the best existing information on EMA and, at the same time, to update it and add to it as necessary. The IFAC EMA standard (IFAC, 2005) falls into the middle ground between regulatory requirements, standards and pure information. As such, its goal is to reduce some of the international confusion on this important topic by providing a general framework and set of definitions for EMA that is fairly comprehensive and as consistent as possible with other existing, widely used environmental accounting frameworks with which EMA must coexist.

This book is thus based on two previous publications I had the pleasure to write:

1. Environmental Management Accounting, Procedures and Principles, United Nations Division for Sustainable Development, Department of Economic and Social Affairs (United Nations publication, Sales No. 01.II.A.3), New York, 2001, www.un.org/esa/sustdev/estema1.htm and
2. Environmental Management Accounting, International Guidance Document, IFAC, International Federation of Accountants, New York, August 2005, www.ifac.org

But, in addition to these two publications this book reflects experiences gained in several case studies and is therefore a true workbook intended to assist organizations in linking their information systems and securing consistency of data for internal management decisions as well as for external reporting purposes. The excel templates used for the assessment of annual environmental costs of organizations, to which I refer to in this book, are available for download at the IÖW's webpage under www.ioew.at and at Springer's homepage under www.springer.com/978-1-4020-9027-1. I hope that with this aid several organizations will be assisted in installing corporate wide internal standards for the collection of material flow and environmental cost data and thereby realizing win-win situations for their bottom line as well as for their environmental performance.

Several current activities emphasize the growing relevance of EMA:

- In March 2008 the International Standardization Organization (ISO) accepted Material Flow Cost Accounting as a new work item within the ISO 14000 Environmental Management Standards series. The Work Item Proposal is explicitly based on the two previously mentioned publications.
- The London Group on Environmental Accounting on request by the UN Committee of Experts on Environmental-Economic Accounting is currently revising SEEA-2003, the worldwide handbook of national environmental-economic accounting (UN SEEA 2003). Consistency with the terminology and concepts of EMA as developed by the UN EMA Working Group is one of the issues on their agenda.
- The Global Reporting Initiative (GRI) released its third version of the GRI Guidelines for Sustainability Performance Indicators and Reporting in winter 2006. Indicator 30 on environmental expenditure directly refers to the IFAC EMA guidance document for the definitions and description in the indicator assessment protocol.

The objective of this book is to define principles and procedures for Environmental Management Accounting (EMA), with a focus on techniques for quantifying environmental costs and material flow data, as a basis for the development of internal EMA assessment guidelines and procedures. The intended users are environmental, production and financial departments of manufacturing companies as well as management consultants and accountants interested in establishing EMA guidelines to support environmental management systems and for better controlling and benchmarking purposes.

Chapter 1 discusses the terms, range and relevance of environmental costs and environmental accounting. Section 1.1 briefly reviews why organizations should care about environmental issues and who should be involved in the set up of an EMA system. It gives an overview on the issues addressed in this book and the fundamental concept of integrated pollution prevention and improved energy and material efficiency on which EMA is based. At the same time there is increasing recognition of the potential monetary benefits of improved environmental performance.

Section 1.2 relates the information needs for environmental performance monitoring with challenges for current accounting practices: such as inadequate communication between accounting and production departments, missing links between the production planning and financial information systems, the prevailing practice of hiding environmental costs in overhead accounts, inadequate tracking of information on materials use, flows and costs, problematic posting of inventory differences and resultantly investment appraisal decisions being based on incomplete information.

Section 1.3 provides definitions for Environmental Costs, Environmental Accounting (EA) and Environmental Management Accounting (EMA). EA is a broad term found in a number of different accounting contexts: financial accounting and reporting; management accounting; externalities estimation (such as full cost accounting); natural resource accounting, national accounting and reporting, and sustainability accounting. The EMA definition of the EMA Expert Working Group of the United Nations Division of Sustainable Development (UNDSD) specifically highlights the two types of information typically considered under EMA: physical and monetary information.

Section 1.4 briefly outlines the terminology of accounting concepts and distinguishes between management accounting (MA), which focuses on internal decision making, and financial accounting (FA), which provides information to external stakeholders.

Section 1.5 describes the physical accounting part of EMA. Physical information includes data on the use, flows and final destiny of energy, water, materials and wastes. EMA places a particular emphasis on physical information because (1) the use of energy, water and materials, as well as the generation of waste and emissions, are directly related to many of the environmental impacts of organizational operations and (2) materials purchase costs are a major cost driver in many organizations.

Section 1.6 relates EMA to financial, statistical, environmental and sustainability reporting requirements. It specifically focuses on highlighting the differences between the EMA approach and the approach taken by SEEA, the System of Environmental Economic Accounting of the United Nations developed for statistical agencies, in the definition of environmental investments and expenditure. The chapter also provides the related requirements of the Global Reporting Initiative (GRI).

Section 1.7 explores EMA uses and benefits. The main areas of application of EMA are internal calculations and decision making. It is however often external pressure that is forcing organizations to look for creative and cost-efficient ways to manage and minimize environmental impacts. EMA can be implemented for different system boundaries, from the corporation to specific processes. From an accountant's point of view, the most likely starting point for EMA is the list of accounts, which is the most common source of cost information in all organizations. From an environmental manager's point of view, the desired starting point may be the analysis of a particular waste stream. A production manager might be the most interested in monitoring a particular product line or set of production equipment.

Chapter 2 describes the input side of the material flow balance. The physical accounting information collected under EMA is a prerequisite for the calculation of many environment-related costs. Mass balances in volumes, energy content and liters and materials flow accounting in monetary terms are the basis for EMA assessments. The physical categories are in line with the general structure of ISO 14031 for environmental performance indicators for the operational system. These physical categories may be subdivided as needed to suit specific business sectors or individual organizations.

Inputs are any energy, water or other materials that enter an organization. Outputs are any products, wastes or other materials that leave an organization. Any Output that is not a Product Output is by definition a Non-Product Output (NPO). In organizations that use energy and materials but do not manufacture physical products, such as transport or other service sector companies, all energy, water and other materials used will eventually leave as Non-Product Output, by definition. The remainder of this document will use the term NPO synonymously with the term "Waste and Emissions." Table 1 describes the main categories of Inputs and Outputs.

Materials Inputs comprise raw and auxiliary materials, packaging materials, merchandise, operating materials, water and energy. Capital items, such as equipment and buildings, are not monitored via mass balances, but can be tracked separately. The environment-related costs associated with the purchase of equipment and other capital items are covered in Chapter 5, via the inclusion of annual depreciation in the appropriate cost categories.

Chapter 3 describes the output side of the material flow balance, which is assessed only in physical, not monetary terms, as the related costs are traced separately. Product Outputs are products and by-products including their packaging. Non-Product Outputs comprise solid waste, wastewater and air emissions.

Table 1 Overview on the input–output material flow balance

Inputs	Outputs
Raw materials	Product output
Auxiliary materials	Products and by-products
Packaging	Non-product-output (NPO)
Operating materials	Solid waste
Energy	Waste water
Water	Air emissions

Chapter 4 deals with environmental performance indicators, which for the operational system are directly derived from the input output material flow balance. The definitions provided in the ISO 14031 standard as well as the related indicators recommended by the Global Reporting Initiative are described. In addition the chapter discusses requirements and system boundaries for indicator systems and specifically addresses the problem of finding meaningful denominators for performance indicators. The chapter concludes with a case study from the brewery in Murau which calculates savings based on their environmental performance indicator system.

Chapter 5 describes the different types of environmentally relevant equipment, which is often the first step when conducting an EMA assessment. The term "equipment" may comprise a single machine or an entire production hall, but the assessment is best performed on a cost center level. In order to provide the necessary data for investment appraisal, actually three categories of environmentally relevant equipment should be distinguished:

- End-of-pipe equipment for treatment of waste and emissions
- Integrated cleaner technologies which prevent emissions at source
- Scrap producing equipment and energy conversion losses

The different approaches of IFAC, UN DSD and UNIDO in opposition to SEEA and CEPA regarding the inclusion of cleaner technologies and integrated prevention are highlighted.

Chapter 6 describes the different environmental cost categories in detail. They are based on the IFAC EMA Guidance Document and comprise the categories described in Table 2.

For each cost category the sub-categories relating to financial accounts, such as equipment depreciation, operating materials, water, energy and personnel are discussed and examples provided. In addition, environment related earnings from grants for investments or from scrap sales are described. National statistical institutes require reporting of environmental costs by the environmental domain affected. The chapter concludes with a case study of the pulp and paper company SCA Laakirchen, which shows the average percentage distribution of the previously described environmental cost categories.

Chapter 7 focuses on linking the physical and monetary information system. It starts with consistency and consolidation issues to be considered when defining

Table 2 Overview on IFACs environmental cost categories

1. Materials costs of product outputs
2. Materials costs of non-product outputs
3. Waste and emission control costs
4. Prevention and other environmental management costs
5. Research and development costs
6. Less tangible costs

the system boundaries for an EMA assessment and when aggregating data from several sites or companies. The chapter deals with information available on the company level, traces environmental aspects in the balance sheet and where to find them in the profit and loss accounts. Chapter 7 goes one step further down into the organization and highlights the principles and terminology of cost accounting, process flow charts and overhead cost attribution. The concepts of activity based costing and material flow cost accounting are explained as well as where to get the necessary data from stock management and production planning systems. The last issues dealt with are application for investment appraisal, budgeting and benchmarking. Danisco, a global supplier to the food industry, uses EMA as a tool primarily to benchmark production sites, which are divers from a geographical and production process point of view in order to demonstrate differences and similarities.

Chapter 8 describes a case study developed from the brewery Murau in depth and at the same time demonstrates how to use the excel template for the EMA cost assessment that is provided as a download under www.ioew.at and at Springer's homepage under. www.springer.com/978-1-4020-9027-1

Chapter 9 describes how to organize an EMA pilot project. The competencies of the project team, selection of sites for pilot testing and a general project plan are discussed. The result of such an EMA pilot assessment may be a company specific adoption of the excel template with more specific cost categories and predefined sources of information as well as an internal procedure which specifies roles and responsibilities. Extracting EMA data from Enterprise Resource Planning Systems and possible elements of an internal EMA assessment standard are explained based on experiences of case studies with Verbundgesellschaft, OMV and Petrom. The chapter ends with a summary of recommendations from about 50 case studies performed so far. The outlook tries to analyze, why companies have been so slow in adopting EMA and MFCA since there is little merit in two separate information systems in an organization, one for financial and cost accounting, the other for process technicians, if "in principle" they should be the same, following the material flows through the company.

The **Annex** provides checklists for environmentally relevant equipment and environmental costs by environmental domains.

Chapter 1
What Is EMA and Why Is It Relevant?

Chapter 1 discusses the terms, range and relevance of environmental costs and environmental accounting. Section 1.1 briefly reviews why organizations should care about environmental issues and who should be involved in the set up of an EMA system. Section 1.2 relates the information needs for environmental performance monitoring with challenges for current accounting practices. Section 1.3 provides definitions for Environmental Costs, Environmental Accounting (EA) and Environmental Management Accounting (EMA). Section 1.4 briefly outlines the terminology of accounting concepts. Section 1.5 describes the physical accounting part of EMA. Section 1.6 relates EMA to financial, statistical, environmental and sustainability reporting requirements. Section 1.7 explores EMA uses and benefits.

1.1 The Issues Behind EMA

The objective of this book is to define principles and procedures for Environmental Management Accounting (EMA), with a focus on techniques for quantifying environmental and material flow costs based on accounting information systems and on developing company internal guidelines for consistent and auditable EMA assessments and data.

The intended users of these EMA principles and procedures are accountants, environmental and production managers interested in installing corporate EMA guidelines appropriate to their own organizations. Such an EMA system will be useful for better controlling and benchmarking purposes and facilitate several external disclosure needs. It is thus also of interest for persons in charge of developing disclosure requirements (such as statistical and other national agencies), auditing the data submitted and consulting on the establishment of such an information system.

Accounting is done in monetary and physical units, but the two are often not consistently linked together. Accountants have a special role in EMA, or certainly should have, since they are the ones with access to the monetary data and information

C. Jasch, *Environmental and Material Flow Cost Accounting*: Principles and Procedures, Eco-Efficiency in Industry and Science 25,

systems needed for EMA activities, the ability to improve or verify the quality of such information and the skills to use that information for decision making.

Experience shows that the environmental manager barely has access to the actual accounting documents of an organization and is only aware of a tiny fraction of aggregate environmental costs. Production departments often keep their own records on the physical inputs and flows related with production. On the other hand, the accountant does have most of the information but is unable to separate the environmental part from the framework of existing accounts without further guidance (Institute of Chartered Accountants 1996 & 2004, Howes, 2002). In addition, the technical and financial departments tend to have communication difficulties, as they use different “languages”.

The limits of traditional financial and cost accounting methods to reflect organizations’ efforts towards sustainability and to provide management with information needed to make sustainable business decisions have broadly been recognized (e.g. Bennett et al. 1998, Burritt et al. 2002, Fichter et al. 1997, German Federal Ministry 2003, Gray et al. 2001, Japanese Ministry of the Environment, 2002, Japanese Ministry of Economy, Trade and Industry, 2002, Schaltegger 2000 & 1996, UNCTAD 1999). Information on environmental performance of organizations might be available to some extend, but, company internal as well as in public authorities, decision-makers are seldom able to link environmental information to economic variables and are crucially lacking environmental cost information.

This is partly due to the definitions applied by statistical agencies in their attempt to capture only “additional” environmental expenditure and investments, which don’t pay back because of efficiency increases.

As a consequence, decision-makers on a micro and macro level fail to recognize the economic value of natural resources, material and energy efficiency improvements and the financial value of good environmental performance. However, sharply rising energy and resource prices, the climate change policy mix and verification of CO_2 emissions have all contributed to the rising necessity of linking material and energy flow data in physical terms with financial information.

Although differing definitions and applications exist, the general use of EMA information is for internal organizational calculation and decision-making (UNDSD, 2001). EMA information for internal decision-making includes both: physical data for material and energy consumption, flows, and final disposal, and monetarised data for costs, savings, and revenues related to activities with a potential environmental impact. The data most useful for decision-making depends on the type of organization (e.g. manufacturing vs. service sector) and the types of decisions to be made (e.g., purchase decisions about raw materials; investment decisions for energy efficiency improvements; altered product design to reduce environmental impact).

EMA data support environmental management systems like ISO 14001 and decision making with regard to improvement targets and investment options. Linked financial and environmental performance indicators are important for controlling and benchmarking purposes. The material flow balances as well as the derived indicators are vital information for environmental reporting. Ranking agencies are interested to see combined monetary and physical approaches towards sustainability, cleaner production and pollution prevention.

The costs for industry of environmental protection, including pollution reduction, waste management, monitoring, regulatory reporting, legal fees and insurance, have increased rapidly in the past 30 years with increasingly stringent environmental regulations. Conventional management accounting systems attribute many of those environmental costs to general overhead accounts, with the consequence that product and production managers have no incentive to reduce environmental costs and are often unaware of the extent of environmental and material flow related costs.

In conventional cost accounting, the aggregation of costs to overhead accounts instead of production cost centers results in their being "hidden" from management. There is substantial evidence that management tends to underestimate the extent and growth of such costs. When environmental costs are allocated to overhead accounts shared by all product lines, products with low environmental costs subsidize those with high costs. This results in incorrect product pricing which reduces profitability.

By identifying, assessing and allocating environmental and material flow costs, EMA allows management to identify opportunities for cost savings. Examples are the savings that can result from replacement of materials that result in hazardous waste, thus eliminating the growing costs of regulatory reporting e.g. under the European REACH Directive, hazardous waste handling and other costs associated with the use of toxic materials and other chemicals. Many other examples (e.g. Envirowise 2003) deal with more efficient material use, highlighting the fact that waste is expensive not because of disposal fees, but because of the wasted material purchase value. Waste and emissions therefore is a sign of inefficient production.

The definition of the "environmental" part of these costs is often troublesome. As well as for cleaner technologies, which are often more efficient in many aspects and prevent the emissions at source, as for many other costs, which often include increased efficiency or health and safety aspects, the "environmental" part is not a precise share. To the extreme, on can say, that if a solution is 100% for the "environment", it often is actually not, because then it will most likely be an end-of-the-pipe treatment, which doesn't solve the problem at source, but shifts it to another environmental media (e.g. dust filters which reduce emissions to air by capturing components which are washed out by rainwater when the filters are disposed of on landfill). These approaches are costly and not efficient.

UNIDOs webpage (www.unido.org/cp) defines cleaner production as a preventive, integrated strategy that is applied to the entire production cycle to

- Increase productivity by ensuring a more efficient use of raw materials, energy and water
- Promote better environmental performance through reduction at source of waste and emissions
- Reduce the environmental impact of products throughout their life cycle by the design of environmentally friendly but cost-effective products

Because of the integrated preventive environmental strategy to processes, products, and services to increase overall efficiency, and reduce risks to humans and the environment the cost related with Cleaner Production and Pollution Prevention can no longer simply be traced from a few clean up technologies and disposal costs. The related activities need to be clearly defined and monitored.

The approach presented in this book has the underlying assumption, that all purchased materials must by physical necessity leave the company either as product or waste and emission. Waste is thus a sign of inefficient production. Therefore when calculating environmental costs, not only environmental protection measures and disposal fees are regarded, but the wasted material purchase value and the production costs of waste and emissions are calculated (material flow cost accounting).

A relatively simple application of EMA that may yield large cost savings is waste management, as the costs of handling and disposing of waste are relatively easy to define and to allocate to specific production steps and products. Enhancing efficiency in the use of energy, water and other raw materials reduces not only environmental impact (reduced resource use and reduced waste and emissions), but also brings about potentially significant monetary savings as the costs of materials purchase and waste treatment decrease accordingly. Other environmental costs, including costs of regulatory compliance, legal costs, damage to the corporate image, and environmental liabilities and risks, are more difficult to assess. But, the largest part of all environmental costs in all manufacturing companies is the material purchase value of non-product output and can be 100 times higher than the costs of disposal, depending on the business sector and the country where the site is situated.

Financial accounts include most of these costs, but aggregated in a way that does not identify the specifically environmental costs and material purchase losses. There is evidence, however, that some environmental liabilities and risks that are in principle covered by reporting requirements are often not reported, for example liabilities for cleaning up of contaminated land. A comprehensive EMA system would promote more complete financial accounts in such cases.

Still, future costs and less tangible costs are hardly found in the existing accounting records. The expected future costs for a necessary wastewater treatment plant upgrade should be part of the current budgeting cycle. Less tangible costs like potential future liability claims and company image costs from poor environmental performance should be considered when comparing investment options.

However, for production companies the most significant costs occur in relation with the materials lost in waste and emissions. Adding the purchase value of non-material output to the environmental costs, makes the share of "environmental" costs higher in relation to other costs. However, it is not the goal of this book to show, that environmental protection is expensive. The costs of materials lost as waste and emissions are not considered part of "environmental protection", but are necessary information for environmental management and investment appraisal of cleaner technologies. It is also not the most important task to spend a lot of time defining exactly which costs are "environmental" or which costs are not, or what percentage of something is "environmental" or not.

The most important task is to make sure that all relevant, significant costs are considered when making business decisions. In other words, "environmental" costs are just a subset of the bigger cost universe that is necessary for good decision making. "Environmental" costs are part of an integrated system of material and money flows throughout a corporation, and not a separate type of cost altogether. Doing EMA is simply doing better, more comprehensive management accounting, while wearing an "environmental" hat, which opens the eyes for hidden costs. Therefore, the focus of material flow accounting is no longer on assessing the total "environmental" costs, but on a revised calculation of production costs on the basis of material flows and fates.

1.2 Challenges for Current Accounting Practices

This section starts with a short introduction to common accounting concepts and language, both for accountants in countries that may have different accounting languages and practices, as well as for any non-accountant readers who may not be familiar with accounting terminology at all.

The two broad categories of accounting that typically take place within an organization are management or cost accounting (MA) and financial accounting (FA). In general, FA tends to refer to accounting activities and the preparation of financial statements directed to external stakeholders, while MA focuses on providing information to organizational management for internal decision making. The two are however closely related and many organizations apply only one system for both purposes. Bookkeeping is the data collection process that generates information for both MA and FA. Total costs and earnings calculated for MA purposes directly are related to the organization-wide revenues and expenditures collected for FA when assessing production costs and product prices.

Financial Accounting is mainly designed to satisfy the information needs of external stakeholders, such as investors, tax authorities and creditors, all of whom have a strong interest in receiving accurate, standardized information about an organization's financial performance. Financial reporting is regulated by national laws and international standards, which specify in detail how different financial items should be treated. The reason behind is to ensure that financial statements are compatible and that taxes are levied on the same basis.

FA relies on standardized financial information. Information on annual revenues and expenditures is provided in an Income Statement (also called Income–Expenditure Account or Profit–Loss Account). The Balance Sheet reports assets, liabilities and equity at a specified date. In addition, the financial statements include a Cash Flow Statement. In addition to data collection and account balancing, FA activities include auditing of the financial statements by financial authorities and for larger organizations by external auditors as well as external reporting.

On the contrary, **Management Accounting** is designed to satisfying the information needs of internal management and provides data for product pricing, investment appraisal and other decision making. Although there are accepted good practices for MA, it is generally not regulated by law. Each organization can determine which MA practices and information are best suited to its organizational goals and needs.

MA focuses on both monetary and non-monetary information (for example, cost drivers such as labor hours and quantities of raw materials purchased) that inform management decisions and activities such as planning and budgeting, ensuring efficient use of resources, performance measurement and formulation of business policy and strategy. MA activities include data collection as well as routine and more strategic analysis of the data via various techniques (such as capital investment appraisal, benchmarking of sites and other controlling activities) designed to address specific management needs.

According to IFAC (1998) the leading-edge practice of MA has shifted in the last years beyond information provision to focus on the reduction of waste (the reduction of resource loss) and the generation of value (the effective use of resources). In other words, MA should focus on the efficient use of resources, which are defined as "monetary and physical" resources, along with the other resources an organization creates and uses, such as "work processes and systems, trained personnel, innovative capacities, morale, flexible cultures, and even committed customers." The role of management accountants in organizations applying this focus has likewise shifted from information tracking to more strategic roles in policy and planning.

Conventional accounting systems and practices have several limitations, which make efficient and consistent data collection regarding environmental and material flow costs a real challenge for production and environmental managers. These limitations can lead to management decisions based on missing, inaccurate and/or misinterpreted information. Especially in investment appraisal the potential future resource costs and benefits of improved environmental performance are often underestimated.

1.2.1 Communication Between Accounting and Production Departments

Accountants and Engineers are trained in different technical language and thus may find it difficult to communicate with each other.

While the environmental manager will have a great deal of knowledge about environmental aspects and impacts of the organization and technical staff will have experience on the flows of materials, energy and water throughout the organization, both environmental and technical personnel often has little knowledge on how these issues are reflected in the accounting system. Often, they also do not have entry permits into the financial subsystems.

As accounting personnel is often unable to provide the information requested by the technical departments out of their system, the environmental and technical departments tend to install additional records in order to trace the data they need. This information may differ quite significantly from the data recorded in the financial departments. It is not unusual to receive quite differing answers to questions regarding the amounts of materials and energy used and total disposal costs from different people.

On the other hand, the accountant or controller has a lot of top down financial information at hand, but often has little knowledge on the actual physical flow of materials and energy through production, the environmental impacts related with them and the environmental relevance of corporate activities.

It is essential for environmental and material flow cost accounting, that accounting, production and environmental management jointly work in a team to assess the data required and install a consistent information system.

1.2.2 Missing Links Between the Production Planning and Financial Information System

Engineering and accounting information systems are often installed as separate satellite systems with system designs that follow completely different logics and thus have no standardized interfaces installed. This may be intentional, as information can be a source of power in organizations.

A good exercise for an internal workshop between production and accounting departments is the mapping of the structure of cost centers with the structure of material and energy flow related information systems in physical terms. It should result in the definition of specified interfaces for consistency checks. Chapter 7.5 deals with this in more detail.

With policy instruments like emission trading systems, this mapping may become mandatory: As the calculation of CO_2 emissions according to the European Emission Trading Scheme is based on verified data for material and energy inputs into relevant production processes, this consistency of data in Austria is verified by a team of external auditors consisting of an engineer, a chemist and an account.

1.2.3 Hiding Environmental Costs in Overhead Accounts

Who is responsible for waste and emissions and related the costs? Different department may have different answers. The production cost centers produce waste and emissions but may have no data on the specific amounts and related costs. The environmental manager does not produce waste, but is in charge of disposing of it. The accounting department may inadvertently "hide" environment related costs by placing them in overhead accounts.

There are numerous examples of potentially important environment-related costs which were hidden on accounting records, where a production manager who would benefit from that information cannot find it easily. One particularly common way to of posting environment-related costs is to assign them to overhead accounts rather than directly to the processes or products that created the costs. While overhead accounts are a convenient way to collect costs that may be difficult to assign directly to processes or products, this practice can create problems later if needed cost information can no longer be traced. It might not be immediately obvious to a manager that an account called "Divisional Overhead" contains information on environmental permit fees, training costs and legal expenses. The posting of potentially significant environment-related costs in overhead accounts may also obscure which are fixed costs that are difficult to reduce and which are variable costs that could be reduced by preventive environmental management.

The posting of environment-related costs to overhead can also be problematic when overhead costs are later allocated back to cost centers (processes, products or services) for pricing and other purposes. Overhead costs typically are allocated back to cost centers by using production related allocation bases, such as production volume, machine or personnel hours. This might, however, be an inaccurate way to allocate some typical environment-related costs. An example would be hazardous waste disposal costs, which might be quite high for a product line that uses hazardous materials and quite low for another that does not. In this case, the allocation of hazardous waste disposal costs on the basis of production volume would be inaccurate, as would be product pricing and other decisions based on that information.

One common approach of resolving this issue is to set up additional cost categories or cost centers for the collection of environment related costs. Often, a cost center for environmental, health and safety management is being installed. But for significant environmental costs a posting to production cost centers or product costs would be preferable. Especially the costs for waste disposal and related material input losses should be posted to the production steps involved and remain in the responsibility of the production managers. This is also promoted by material flow cost accounting.

1.2.4 Posting of Inventory Differences

Although larger production companies annually generate millions of data records concerning material flows from Enterprise Resource Planning (ERP), Production Planning Systems (PPS) and other software systems, the available information is often not sufficiently accurate or detailed for environmental, efficiency and other decision-making purposes. If the system has been installed from a pure financial accounting perspective, the information related to materials inputs, flows, fates and related costs is often not tracked adequately.

Several case studies revealed that the posting of materials purchase information does not allow clear identification of the amount and value of different categories of purchased materials. In some accounting systems, all material purchased is posted on one account, while the detailed material numbers and amounts are recorded only in the stock management records. So, there is no easy way to aggregate the data from stock management by materials group or trace the actual annual consumption of the different categories of materials. A time-consuming and expensive manual process of data reorganization and comparison would be required. Thus, no one knows the amount and value of materials consumed by materials groups neither for the company nor by cost centers.

Even if a production manager has estimates of material loss percentages during the production process, the total value of lost materials often can not be calculated because of missing data on the value of materials purchased by materials groups. As the desired materials purchase information is often difficult to extract from the accounting systems, some environmental managers when installing their environmental management system have asked their materials suppliers to provide this information instead. Although this might be a cost-efficient solution for a specific project, in general, an organization should set up its own data systems to provide the needed information for ongoing materials flow and environmental management.

Another example is the practice of aggregating materials purchase costs and materials processing costs (such as labor) into a single cost account. For a company that uses several manufacturing steps to make its final product, the value of the semi-finished product entering the final manufacturing step is accurately viewed as the sum of all costs of materials purchase and processing incorporated into that semi-finished product. If, however, this cost information is recorded in the accounting records as a single lump sum figure, with no detail on the split between materials purchase costs and other processing costs, the disaggregation of these costs for later decision making can be difficult and time consuming.

In addition, conventional cost accounting systems often do not record data on material inputs to and from each cost center in production, but rely on general calculations provided by the production planning system, which may or may not reflect an organization's real-world use and flow of materials. Many production-planning systems calculate materials loss by using inaccurate average loss percentages. They may have little to do with the actual losses that occur during production. The employees on-site often have more precise estimates than the accounting system does. Chapter 7 will deal in depth with these issues.

1.2.5 Investment Appraisal Based on Incomplete Information

As has been shown, environmental costs are often not adequately monitored. Of course, this is also true for related earnings and cost savings. In addition, most of these costs are usually not traced systematically and attributed to the responsible processes and products, but simply summed up in general overhead.

The fact that environmental costs are not fully recorded often leads to distorted calculations for improvement options. Environment protection projects, aiming to prevent emissions and waste at the source (avoidance option) by better utilizing raw and auxiliary materials and requiring less (harmful) operating materials are not recognized and implemented. The economic and ecological advantages of such measures are not realized. The people in charge are often not aware that producing waste and emissions is usually more expensive than disposing of them.

Investment appraisal is based on estimates regarding future costs of materials, products and processes. In general, current costs are extrapolated. But, if current costs are incompletely monitored, decisions on investment projects, materials choices, product pricing and product mix suffer. Investment decisions pose particular challenges because they involve the uncertainty of questions such as: What will I have to pay in the future if I do not act now? What will I earn in the future if I do act now? A lack of accurate estimates of environment-related cost and benefits adds to the inherent uncertainty of all investment decisions.

The main problem associated with a systematic identification of the potential for material efficiency improvements lies in the traditional cost accounting systems which are not able to provide the relevant information on the company's physical structure, i.e. on the structure of its material flows. In particular the non-product output (waste, wastewater, etc.) is not being quantified and monetarised separately within accounting systems.

Organizations need to consider all potentially significant environment-related costs that may influence the return on investment, such as materials flow costs, site recovery costs and any costs associated with certain or likely future regulations. Organizations also need to ensure that environmental managers, technical experts and accountants work together in providing the full picture of environmental issues and the related costs and benefits that are relevant for making an investment decision.

1.3 Definition of Environmental Costs and Environmental Management Accounting (EMA)

From a macroeconomic perspective, the prices for scarce raw materials, pollution and disposal do not reflect their true value and cost to society. Health hazards, repairs of contaminated sites etc. are environmental costs usually not borne by the polluter but by the general public. **Environmental costs** comprise both internal and external costs and relate to all costs occurred in relation with environmental damage and protection. **Environmental protection costs** include costs for prevention, disposal, planning, control, shifting actions and damage repair that can occur at companies, governments or people (Association of German Engineers, 2001).

The focus of EMA is on **corporate** environmental costs. External costs which result from corporate activities but are not internalized via regulations and prices are not considered for the assessment of current costs, but may be an issue for investment appraisal. It is the role of governments to apply political instruments such as eco-taxes and emission control regulations in order to enforce the 'polluter-pays' principle and thus to integrate external costs into corporate calculations. The methods to assess these costs are summarized under the term EA (instead of EMA).

Environmental Accounting (EA) is a broad term used in a number of different contexts, such as (IFAC, 2005):

- Assessment and disclosure of environment-related financial information in the context of financial accounting and reporting
- Assessment and use of environment-related physical and monetary information in the context of Environmental Management Accounting (EMA)
- Estimation of external environmental impacts and costs, often referred to as Full Cost Accounting (FCA) (Bebbington et al. 2001, Canadian Institute of Chartered Accountants 1997)
- Accounting for stocks and flows of natural resources in both physical and monetary terms, that is, Natural Resource Accounting (NRA)

- Aggregation and reporting of organization-level accounting information, natural resource accounting information and other information for national accounting purposes and
- Consideration of environment-related physical and monetary information in the broader context of sustainability accounting

What then are **corporate environmental costs**? Costs incurred to deal with contaminated sites, effluent control technologies and waste disposal may first come to mind. They have impact both on **management accounting** (assessment of an organization's costs for pollution control equipment; earnings from recycled materials; annual monetary savings from new energy-efficient equipment) and **financial accounting** (evaluation and reporting of the organization's current environment-related liabilities).

The next step is to define environmental protection. **Measures for environmental protection** comprise all activities taken for legal compliance, compliance with own commitments or voluntarily. Economic effects are no criteria, but the effect on prevention or reduction of environmental impact (Association of German Engineers, 2001).

Corporate environmental protection expenditure includes all expenditure for measures for environmental protection of a company or on its behalf to prevent, reduce, control and document environmental aspects, impacts and hazards, as well as disposal, treatment, sanitation and clean up expenditure. The amount of corporate environmental protection expenditure is not directly related to the environmental performance of a company (Association of German Engineers, 2001).

The Association's of German Engineers definitions for environmental protection comprise both **prevention and treatment** activities. But for company internal calculation of environment-related costs, expenditure for environmental protection is only one part of the coin. The costs of waste and emissions include much more then the respective treatment facilities and disposal fees.

The concept of '**waste**' has a double meaning. Waste is a material which has been purchased and paid for, but which has not turned into a marketable product. Waste is therefore indicative of production inefficiency. For the assessment of total annual environmental costs as a basis for future calculations and decisions, the costs of wasted materials, capital and labor have to be added. Waste in this context is used as general term for solid waste, waste water and air emissions, and thus comprises all **non-product output**. **Materials** include water and energy.

The approach developed for the UN CSD (Jasch, 2001) assumes that all purchased materials leave the company either as a product or as emissions and waste (unless stored) (Table 1.1).

Table 1.1 Total corporate environmental costs

	Environmental protection expenditure (end-of-pipe emissions treatment and integrated waste prevention)
+	Material flow costs (costs of unproductive material, capital, and personnel)
=	Total corporate environmental costs

From a business perspective, it thus makes sense to minimize (environmental) costs, but not because of abandoning environmental protection, but because of integrated production processes which don't produce waste and don't require emission treatment. This makes sense from a micro and well as macro economic perspective.

Environmental costs under EMA include not only Environmental Protection Expenditure, but also other important monetary information needed to cost-effectively manage environmental performance. **Material Flow Costs** comprise the purchase cost of materials that eventually become waste or emissions. The related capital and personal costs to produce waste and emissions may be added, thus calculating production costs of waste. The physical accounting side of material flow cost accounting (MFCA) provides the needed information on the amounts and flows of energy, water, materials and wastes to assess these costs.

Several projects in the manufacturing sector have shown that the costs of waste disposal are typically 1–10% of total environmental costs, while the purchase costs of the wasted materials represent 40–70% of environmental costs depending on the business sector examined.

Material flows are money flows and can therefore in principle be mostly traced by conventional accounting systems. Also, when calculating investments for environmental protection, increased material and production efficiency need consideration (Fig. 1.1).

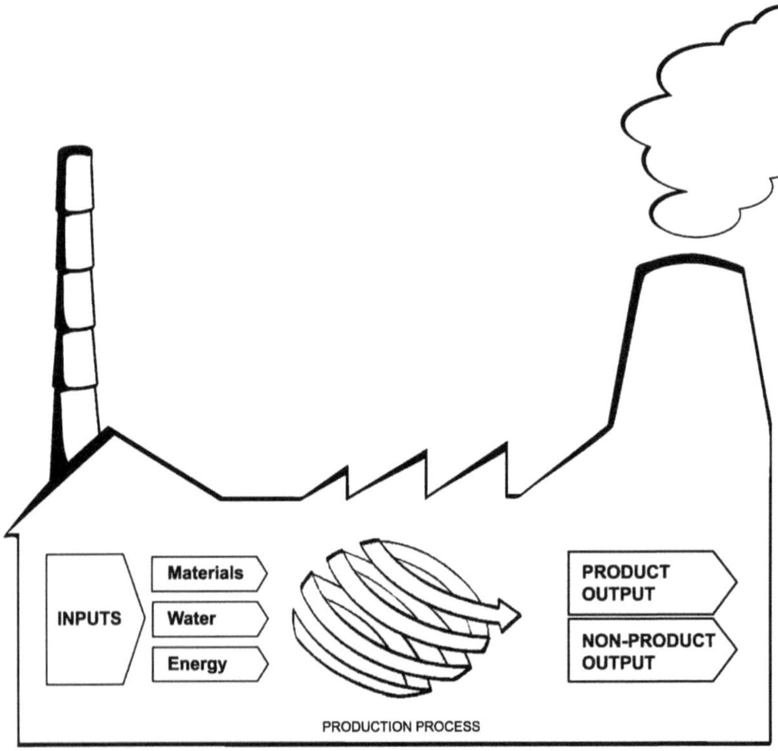

Fig. 1.1 Material flows are money flows

According to the definition of UN DSD (Jasch, 2001) two types of information are considered under EMA: physical and monetary information. Physical information includes data on the use, flows and final destiny of energy, water, materials and wastes. EMA places a particular emphasis on physical information because

1. The use of energy, water and materials, as well as the generation of waste and emissions, are directly related to many of the environmental impacts of organizational operations.
2. Materials purchase costs are a major cost driver in many organizations.

The United Nations Expert Working Group on EMA, which distinctively highlights both the physical and monetary sides of EMA, has developed the following definition for EMA. According to the UN group:

EMA is broadly defined to be the identification, collection, analysis and use of two types of information for internal decision making:

- Physical information on the use, flows and destinies of energy, water and materials (including wastes) and
- Monetary information on environment-related costs, earnings and savings

Under the **physical accounting side of EMA**, an organization should try to track all physical inputs and outputs and ensure that no significant amounts of energy, water or other materials are unaccounted for. The accounting for all energy, water, materials and wastes flowing into and out of an organization is called a "materials balance," sometimes also referred to as "input-output balance," a "mass balance", "material flow balance" or an "eco-balance." (United Nations Environment Program and United Nations Industrial Development Organization, 1991; German Environmental Protection Agency/German Environment Ministry, 1995; Pojasek, 1997; Environmental Protection Agency of Baden-Würthemberg, 1999).

Many organizations perform energy balances and water balances separately from other materials balances. As this terminology implies, the underlying assumption is that all physical inputs must eventually become outputs—either physical products or waste and emissions—and the inputs and outputs must balance. The level of precision of a materials balance can vary, depending on the specific purposes of the information collection and the availability and quality of the data.

Materials Inputs are any energy, water or other materials that enter an organization. Outputs are any products, wastes or other materials that leave an organization. Any Output that is not a Product Output is by definition a Non-Product Output (NPO). In organizations that use energy and materials but do not manufacture physical products, such as transport or other service sector companies, all energy, water and other materials used will eventually leave as Non-Product Output, by definition.

The IFAC guidance document on EMA uses the term NPO synonymously with the term "Waste and Emissions." The Japanese guide for Material Flow Cost Accounting is based on the same concept and distinguishes output into positive and negative products (METI, 2007). The physical categories described by IFAC are also in line with the general structure of ISO 14031 (ISO, 2000) for environmental performance indicators for operational systems (ISO 14031),

Table 1.2 IFAC cost categories for EMA

1. Materials Costs of Product Outputs Includes the *purchase costs* of natural resources such as water and other materials that are converted into products, by-products and packaging.
2. Materials Costs of Non-product Outputs Includes the *purchase (and sometimes processing) costs* of energy, water and other materials that become Non-Product Output (Waste and Emissions).
3. Waste and Emission Control Costs Includes costs for: *handling, treatment and disposal* of Waste and Emissions; *remediation and compensation* costs related to environmental damage; and any control-related *regulatory compliance* costs.
4. Prevention and Other Environmental Management Costs Includes the costs of *preventive environmental management activities* such as cleaner production projects. Also includes costs for *other environmental management activities* such as environmental planning and systems, environmental measurement, environmental communication and any other relevant activities.
5. Research and Development Costs Includes the costs for *Research and Development* projects related to environmental issues.
6. Less Tangible Costs Includes *both internal and external* costs related to less tangible issues. Examples include *liability, future regulations, productivity, company image, stakeholder relations and externalities.*

which are referenced in ISO 14001 (ISO, 1996), the standard for environmental management systems.

For the **monetary accounting side of EMA**, cost definitions from a variety of international sources were reviewed for the IFAC EMA guidance document and a set of cost categories was developed. The goal was to develop a set of cost categories that represents not only widely accepted international practice, but also emerging best practice. Table 1.2 shows the environment-related EMA cost categories of IFAC. For the EMA assessments in addition earnings from investment grants, subsidies and sale of waste for recycling are being recorded.
The IFAC environmental cost categories comprise:

The first cost category, **Materials Costs of Product Output**, reflects the view to regard the purchase costs of all natural resources (energy, water, materials) as environment related. In production companies, where most of the purchased materials are converted into physical products, this allows more cost-effective management of the materials-related environmental impacts of those products and directly relates to the input-output balance of material flows. Of course, organizations do consider materials purchase costs in their internal management decision making, but do not necessarily view them as environment related. These costs can be viewed as environment related, because an organization must have this information to fully assess the financial aspects of environmental management related to both physical waste and physical products. These physical flows are being monitored within environmental management systems and directly relate to improving environmental performance indicators. If properly installed, the monetary accounting side of EMA can provide much of the data needed for the physical accounting side of EMA related with information on the amounts and flows of energy, water, materials and wastes.

The second cost category, **Materials Costs of Non-Product Output**, is also based on the physical material flow balance. For each material group on the input side, the loss percentage is estimated or monitored. Wasted materials are evaluated with their material purchase value or materials consumed value in case of stock management. Technical process flow balances and material flow costing help to assess non-product output more precisely and allow distributing the related costs back to the responsible polluting cost center or cost carrier (product). The **production costs of non-product output** may be calculated with the respective production cost pro rata charges, which include labor hours, depreciation of machinery and operating materials and financing costs.

The third cost category, **Waste and Emission Control Costs**, comprises **disposal and emission treatment costs** including related equipment, labor and maintenance materials. Insurance and provisions for environmental liabilities and clean up also reflect the spirit of treatment instead of prevention. This category corresponds to the conventional definition of environmental costs comprising all treatment, disposal and clean-up costs of existing waste and emissions.

Waste and emission treatment using end-of-pipe technologies is usually the first step on the environmental protection path. End-of-pipe investments are gradually implemented as the need for legal compliance increases. Public as well as corporate activities aimed at environmental management are often still focusing on end-of-pipe technologies, which may in the short run appear to be a fast solution, but in the long run often actually amount to more consumption of material and energy, more capital expenditure and more work hours than if measures are taken at the source.

The forth category, **Prevention and other Environmental Management Costs**, records the labor costs and external services for good housekeeping as well as the "environmental" share of cleaner technologies, if significant. Integrated prevention activities are actually inherent to production and thus the "environmental share" of these costs has to be estimated based on environmental impact reduced, in relation to "standard" production equipment or based on the motives for the expenses.

Pollution prevention can be achieved by two ways, by changes in product design or production processes and by better housekeeping assisted by environmental management systems, with the two approaches often being interlinked. Integrated environmental protection attempts to avoid waste and emissions altogether. Cleaner technologies avoid the need for hazardous operating materials which require costly disposal methods. In contrast to expensive end-of-pipe investments, pollution prevention often significantly reduces environmental costs.

Research and development costs for environmental projects may also be seen as part of pollution prevention. But as national statistical agencies tend to request this data separately, it has also been defined as a stand alone cost category.

The last cost category, **Less Tangible Costs**, deals with costs that are not directly traceable from the accounting system. Examples include *liability, future regulations, productivity, company image, stakeholder relations and externalities*. These potential cost should be especially considered for investment appraisal.

1.4 Monetary Accounting

As not all readers of this book will have an accounting background, but rather a more technical training, this chapter deals with accounting basics. Conventional corporate monetary accounting comprises

- Financial accounting (bookkeeping, balancing, consolidation, auditing of the financial statement and reporting)
- Cost accounting (also called management accounting)
- Corporate statistics and indicators (past oriented)
- Budgeting (future oriented)
- Investment appraisal (future oriented)

Book keeping and cost accounting provide the data basis for the other instruments. They can and have also been used to trace expenditures, costs, indicators, investments and savings, due to measures for environmental protection, but not systematically. Corporate application of financial accounting comprises mainly internal calculation tools, but is also used for external reporting to financial authorities, shareholders and the company register. Statistical agencies make use of this information.

Cost accounting or **management accounting** constitutes the central tool for internal management decisions such as product pricing and is not regulated by law. This internal information system deals with the following questions: What are the production costs for different products and what should be the selling price of these products? For determining the inventories of finished goods and work-in-progress for the balance sheet, cost accounting also needs to be done for financial reporting. The main stakeholders in cost accounting are members of different management levels (e.g. executive, site, and product and production managers). For environmental management, the related costs may be traced and allocated to products and cost centers. The appropriate approach will therefore be described in Chapter 7.

Cost accounting is based on data obtained from financial accounting but sometimes uses different values, e.g. repurchasing values for deprecations, average prices for material input or imputed interest. The latter are assessed differently due to the system of **transition from expenditure to costs**. Most small and medium sized companies use the same figures with only minor adjustments.

Alas, many companies do not have a separate cost accounting system, but calculate on the basis of the financial accounting data instead. **Financial accounting**, on the contrary, is mainly designed to satisfy the information needs of external shareholders and financial authorities, both of whom have a strong economic interest in standardized comparable data and in receiving true and fair information about the actual economic performance of the company. Therefore, financial accounting and reporting are being dealt with in national laws and international accounting standards. They regulate how specific items should be treated, specifying, e.g., whether environmental investments should be capitalized or expensed, under which circumstances provisions may be made for future treatment liabilities, or when contingent liabilities should be disclosed. Imputed (calculatory) approaches as used

Table 1.3 Terminology of financial accounting and cost accounting

Financial accounting	Cost accounting
Balance sheet	
Assets	No equivalent
Liabilities	No equivalent
Profit and loss accounts	**Cost statements**
Expenditures	Costs
Expenditure categories	Cost categories
Revenues	Earnings
No equivalent	Cost centres
Calculation of production expenditure	Cost carriers/objects (Products)

in cost accounting are not permissible. All costs must therefore be recalculated to show actual expenditure and prices.

Financial accounting deals with revenues and expenditures as shown in the profit and loss account, and with assets and liabilities as listed in the balance sheet. More detailed information is available from the list of balances. In cost accounting, the terms dealt with are costs and earnings; there is no equivalent to the balance sheet.

Requiring a somewhat different assessment method, the various expenditure subcategories in financial accounting correspond to the categories of costs which are allocated to the respective cost centers (in-house production processes) and cost carriers/objects (products).

Data determination for the two accounting methods may differ slightly. For financial accounting, the system boundary is the legal entity and therefore mostly the company fence, sometimes, aggregating over several production sites. Cost accounting steps further down, inside the company and traces the costs of production steps and products (Table 1.3).

Environmental management accounting thus represents a combined approach which provides for the transition of data from financial accounting and cost accounting to increase material and energy efficiency, reduce environmental impact and risks and reduce costs of environmental protection. In the following, the term expenditure is always used when a precise distinction to implicit cost approaches is necessary. Otherwise, the term cost is used. For the different cost categories of IFACs environmental cost scheme (Table 1.2), guidance is given on where to find them and how to deal with them when expenditures or costs are assessed.

All expenditure should refer to the same reporting period and be derived from the annual list of accounts, which in the first round means a yearly monitoring of total annual environmental expenditure. This does not include external costs and envisaged future price changes, and the scheme for total annual environmental expenditure is not used for the calculation of investment options or project costs and cost savings. Chapter 7.8 deals with these issues separately.

The assessment can be based on expenditures from the profit and loss accounts or on internal cost accounting documents, depending on the structure of internal information systems. It is the task of the company's controller to define the most appropriate data base once the general outline of the approach to be adopted has been defined. This also depends on the entry permits into the corporate information systems of the people in charge.

Since the environmental cost assessment should also be used for uniform reporting procedures it is recommended to refer to actual expenditure quoted in financial accounting but to allocate it to sites, cost centers and products.

1.5 Physical Accounting

This chapter outlines the type of physical information relevant under EMA in more detail and briefly discusses the related concepts of materials balances, materials flow accounting and environmental performance indicators.

The core part of environmental information systems is **material flow balances in physical units** of material, water and energy flows within a defined system boundary. This can be on the corporate level, but also one step further done to cost centers, sites and production processes or even down to machinery's and products. Then, it becomes the task of process technicians and not so much accountants to tackle and trace the necessary data.

On a higher level, material flow balances are calculated for regions and countries, referred to under the term "national resource accounting". Austria, Germany and Japan were the first countries to have consistent material flow balances for their nations, which are provided by the statistical agencies.

On a national level, statistical agencies and economic sciences also strive to estimate total annual environmental expenditure of industry and the costs to the general public due to environmental pollution (so called external costs, as they are not born by the polluting company, but the general public). External costs are part of environmental accounting as well as national resource accounting in material flows, but both are not management accounting (Table 1.4).

With rising costs for environmental compliance, disposal and the need to improve material efficiency in strongly competitive markets, tracking and tracing material flows throughout the company has been the major tool for detecting potential improvements in waste prevention and cleaner production. Likewise, calculating the related environmental costs and distributing them back to the polluting cost center, process or product (polluter pays principle, also in cost accounting) has gained importance for the correct calculation of the profitability of products, processes and production sites.

Table 1.4 System boundaries for material flow balances

Input	System boundaries	Output
	Nations	
Materials ⇒	Regions	⇒ Products
Energy ⇒	Corporations	⇒ Waste
Water ⇒	Processes	⇒ Emissions
	Products	

The material flow balance is an equation based on "what comes in must go out or be stored". In a material flow balance information on both the materials used and the resulting amounts of product, waste and emissions are stated. Inputs and Outputs are measured in physical units in terms of mass (kg, t), liters or energy (MJ, kWh). The purchased input is cross-checked with the amounts produced and sold as well as the resulting waste and emissions. The goal is to improve efficiency of material management both economically and environmentally.

A material flow balance can be made for a few selected materials or processes, or for all materials and wastes of an organization. The aim of process balances is to track materials on their way through the company. The starting point often is the corporate level, as much information is available only for this system boundary. Also, this level is used for disclosure in environmental reports.

Most organizations purchase energy, water and other materials to support their activities. In the production sector most of the purchased materials are converted into products that are delivered to customers. Most production companies also produce waste—materials that were intended to go into final product but became waste instead because of product design issues, operating inefficiencies, quality issues, etc.

Companies including the service sector also purchase energy, water and materials that are never intended to become a physical product but are necessary to manufacture the product (such as water to rinse out production equipment, fuel for transport operations or chemicals to run the waste water treatment plant). Many of these materials eventually become waste streams that must be managed. Companies outside the production sector (for example, agriculture, resource extraction, services, transport, public administration) can also use a significant amount of energy, water and other operating materials to help run their operations, which, as they are not converted into products, by definition eventually end up as waste and emissions.

Not only the materials input into production and service provision will have an environmental impact, but also the physical products (including by-products and packaging) will have environmental impacts during transport, use at the customer and final disposal. Some of the potential environmental impacts of products can be reduced by changes in product design, such as decreasing the volume of paper used in packaging or replacing a physical product with an equivalent service, etc. Some companies have decided to separately record their efforts to reduce the environmental impact of their products also within the prevention oriented measures of their environmental cost accounting system. Other companies are monitoring the environmental impacts of their products along the whole product life cycle, from raw material extraction, production till the final use and disposal.

To effectively manage and reduce the potential cnvironmental impacts of waste and emissions, as well as of any physical products along its life cycle, an organization must have accurate data on the amounts and final destiny of all the energy, water and materials used to support its activities. It needs to know which and how much energy, water and materials are brought in, which become physical products and which become waste and emissions. This physical accounting information does not provide all of the data needed for effectively managing all potential environmental impacts, but is essential information that the accounting function can provide.

Product life cycle assessments (LCA's) comprise two levels. Company internal is the attribution of the process data (e.g. on a cost center level) to the products produced. This is a prerequisite for corporate LCA's. The system boundary for LCA's follows the product throughout its life cycle by adding upstream and downstream life-cycle stages along supply chains. This method, based on material flow thinking, has been incorporated into ISO 14040 (ISO, 1998).

As obvious, LCA's require very good data quality from corporations. In addition, they mostly require data from companies outside the direct sphere of influence and data which can also not be gathered from environmental reports, as most companies produce more than one product in more than one process. At global level more than 100,000 companies now have an ISO 14001 management system, which again has a comprehensive impact on supply chains. At the same time, experience shows that the comparability of performance indicators and the consistency of the financial and technical information systems are very weak and not much data is being disclosed.

The only solution often available to scientists and consultants is to refer to data published by statistical agencies on the level of industry sectors (NACE Codes). The necessity to rely on data from national statistics for LCA's is increasing as globalization of supply chains has a fast growth rate due to fast growing economies like China, India and Brazil etc. A number of databases for LCA based on national economic and environmental statistics are now available. These databases are known as "Input-Output databases" or "IO-databases" for short.

So while LCA in general terms may work on a macro level, linking highly aggregated sector specific information on material flows with environmental impacts, and providing very general information on environmental impacts from production sectors, the link to data collected on a micro level remains weak. The information, that LCA's based on macro data can supply, is relevant e.g. for political decision making for instance related to environmental labeling, but not so much as a decision making tool for companies when it comes to procurement or ecodesign, as the data is not company specific enough.

Environmental management systems, performance indicators and management accounting have their application on a micro level, but this information is not fed back to the macro level. For performance evaluation and product life-cycle assessment (LCA), the production steps and processes covered by the companies or product systems analyzed must be carefully defined so that the production steps covered by an input-output analysis are identical. Figure 1.2 shows the product life-cycle assessment scheme. Data comparison within sites, processes and products requires that the system boundaries of the participants are comparable; otherwise the results will be meaningless.

As mentioned earlier, materials purchase costs are a major cost driver for many organizations. The physical accounting information collected under EMA is, therefore, key to the development of many environment-related costs. The physical accounting and monetary accounting sides of EMA are integrally linked in many ways.

Ideally, the material flow balance can be summed up to show how much of the purchased materials were actually processed into product sold and how much was

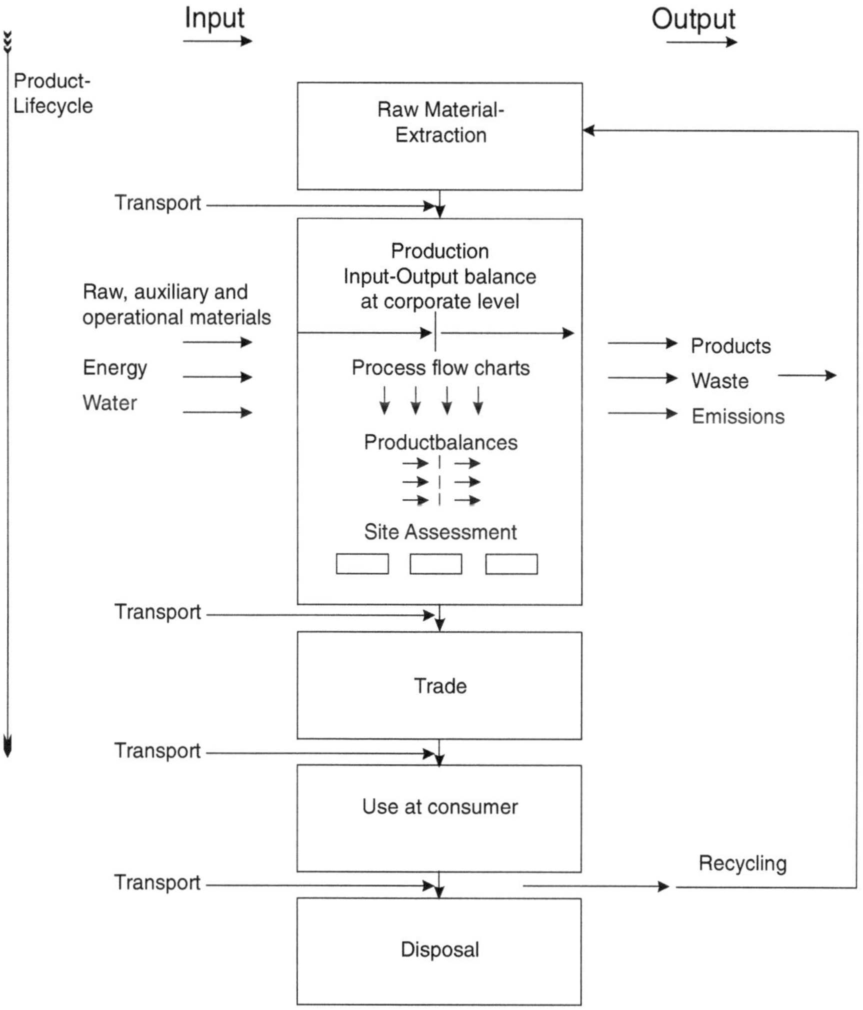

Fig. 1.2 Product life cycle assessment

discharged as waste, waste water or emissions. Figure 1.3 from a PREPARE Pollution Prevention project (Jasch et al., 1997) shows that by **monetary value** only 39% of the raw and auxiliary materials purchased actually left the company as products. The rest ended up in the environment. By **physical volumes** the ratio was even less favorable: only 12% of purchased materials by weight went into the product, the rest had to be disposed of at high costs or had to be treated with cleaning technologies. It is obvious that the disposal costs in this setting account for only the lowest share of environmental costs. It is also obvious that such a production process is less than optimal both from an economic and an environmental point of view.

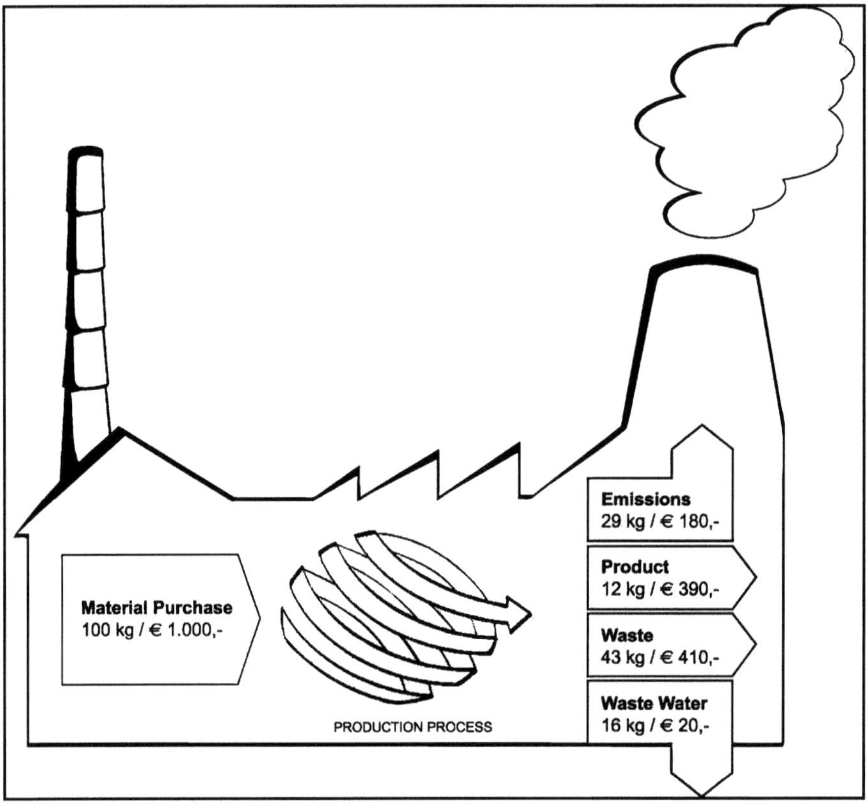

Fig. 1.3 Material and money flows in a paint shop

But, in order to be able to aggregate and evaluate the data for such an analysis, the underlying information system has to provide precise information on the quantities purchased, produced and disposed of. The later chapters will describe how to set up such a system.

During their first environmental review, companies mostly draw up a screening material flow balance and do not go into much detail. On this basis, knowledge is gained on where to focus to achieve improvements in performance and information gathering. By improving the quality of the information available and the consistency of information systems, a regular monitoring system can be established. Such a monitoring system may show resource input, production and waste outputs on a monthly basis. As a next step, the material flows can be subdivided further according to processes and cost centers, and they can then also be subjected to monetary evaluation.

The material flow balance is also the basis for environmental performance indicators. From an environmental impact point of view, the absolute data collected are the most important, as these absolute indicators illustrate the consumption of natural resources and the generation of waste and emissions, such as:

- The total amount of raw materials or energy consumed each year
- The total amount of solid waste or wastewater generated each year

Relative (normalized) indicators represent an organization's environmental performance in terms of its size, production output or number of employees. These are important indicators since company size; product or service output can vary from year to year. Thus, these indicators allow an organization to distinguish between changes in environmental performance as a result of changes in these factors and changes in performance as a result of environmental management efforts. Examples of relative indicators include:

- Amount of raw materials or energy consumed per unit product manufactured or service provided
- Amount of solid waste or wastewater generated per unit product manufactured or service provided

Relative indicators may also tie physical and monetary terms together. Environmental performance indicators are further discussed in Chapter 4.

1.6 EMA Links to Financial, Statistical, Environmental and Sustainability Reporting Requirements

There is a growing trend to include increasing amounts of environment-related financial as well as non-financial information in corporate financial and sustainability reports to external stakeholders. Accountants within organizations play a key role in providing this information, and external auditors play a key role in verifying the accuracy of the information reported, as well as verifying the information systems and practices from which the reported information is derived.

Some companies have published annual environmental investments and annual costs in their environmental reports but it is not immediately obvious if high figures are good or bad as this depends on the type of costs. It is necessary to specify in detail the expenses in the different environmental cost categories as it makes a difference if money is spent on investment or depreciation of End-of-pipe technologies and waste treatment technologies, or if the costs occur for general environmental management and donations for protecting land, or if the majority of environmental costs are the calculated production costs for non-product output.

From a business perspective it is always beneficial to reduce costs, also environmental costs, even if the first reaction might be the impression that less environmental expenditure is less environmental performance.

Sometimes the question asked is: Should we report rather high or low costs?

The answer is: From a financial point of view, lower costs are always better! From an environmental point of view, not the costs, but the environmental impact is important.

It is therefore preferable to invest in technologies and management systems that prevent the creation of waste and emissions at source. But the "environmental share"of these integrated measures is difficult to assess.

From a communications point of view, not the total of environmental costs, but the distribution of environmental costs between Non-product-Output, Waste and Emission Treatment and Prevention as well as the medium term shift from End-of-Pipe to integrated and material flow related measures is of interest and should be communicated.

1.6.1 The EC Recommendation and the EU Directive on Environmental Issues in Company Annual Accounts and Reports

The European Commission adopted a Recommendation on the recognition, measurement and disclosure of environmental issues in the annual accounts and annual reports of companies in May 2001 (European Comission, 2001a). The recommendation covers requirements for recognition, measurement and disclosure of environmental expenditures, environmental liabilities and risks and related assets that arise from transactions and events that affect, or are likely to affect, the financial position and results of the reporting entity. The Recommendation also identifies the type of environmental information that is appropriate to be disclosed in the annual and consolidated accounts and/or the annual and consolidated annual report with regard to the company's attitude towards the environment and the enterprise's environmental performance, to the extent that they may have consequences on the financial position of the company.

The Recommendation states that "Appropriate disclosures are considered a key factor that facilitates transparency of information. Disclosures are appropriate where they affect the user's understanding of the financial statements." The recommendation aims at providing comprehensive guidance in the area of disclosure, and identifies relevant disclosures that allow for comparability and consistency of the environmental information presented. Certain accounting treatments with regard to environmental issues are recommended in order to enhance the provision of more meaningful information by the preparers of the financial statements, with the focus being on treatment of financial liabilities and provisions for clean up and repair.

As such, the definition for environmental expenditure used in the Recommendation is strictly end-of-pipe oriented: "Environmental expenditure includes the costs of steps taken by an undertaking or on its behalf by others to prevent, reduce or repair damage to the environment which results from its operating activities. These costs include, amongst others, the disposal and avoidance of waste, the protection of soil and of surface water and groundwater, the protection of clean air and climate, noise reduction, and the protection of biodiversity and landscape. Only additional identifiable costs that are primarily intended to prevent, reduce or repair damage to the environment should be included. Costs that may influence favorably the environment but whose primary purpose is to respond to other needs, for instance to increase profitability, health and safety at the workplace, safe use of the company's products or production efficiency, should be excluded."

As environmental expenditures are seen solely as additional expenditure due to legal requirements, and there is no link to management accounting, internal cost benefits and investment appraisal logics, resultantly, environmental protection is perceived as an additional burden without any benefit. Efficiency improvements, resulting in less material and energy input and thus reduced emissions at source, are explicitly excluded. The UN DSD concept of EMA, the concepts of integrated pollution prevention and efficiency go much further and highlight the links to internal cost savings, reducing scrap and improving good management practices.

The Recommendation refers to the Classification of Environmental Protection Expenditures (CEPA) developed by Eurostat, the Statistical Office of the European Union. These definitions cover expenditures for activities whose primary purpose is environmental protection. They primarily relate to IFACs Cost Categories (3) Waste and Emission Treatment. The IFAC cost categories (4) Prevention and Other Environmental Management; and (5) Research and Development are in practice mostly excluded because of the exclusion criteria for profitable integrated efficiency measures.

In a workshop at the European Commission in November 2004 the results of a study on the limited application and relevance of the Recommendation were discussed. Several countries and companies presented their more efficiency and prevention oriented approaches to environmental management. The outcome was not to revise the Recommendation, but rather to focus on the application of the related issues in the Modernization Directive, which allows a much more flexible inclusion of issues and performance indicators relevant for environmental protection.

The EU Modernization Directive on the annual and consolidated accounts of certain types of companies, (European Parliament & Council, 2003), stipulates a requirement of inclusion of relevant environmental (and social) information in corporate annual reports. The Directive states that the information in annual reports should not be restricted to the financial aspects of a company's business, but that "To the extent necessary for an understanding of the company's development, performance or position, the analysis shall include both financial, and where appropriate, non-financial key performance indicators relevant to the particular business, including information relating to environmental and employee matters."

Thus, although the original EC Recommendation was voluntary for European countries and companies, the 2003 EU Directive has made the reporting of significant environmental issues and performance indicators in annual accounts and reports mandatory. A current project of the Federation des Experts Comptables, FEE, analyses the degree of national implementation in the European Union and best practice corporate reports. Early results of the study suggest that formally, the requirement of the Modernization Directive has been included in national corporate accounting law, but has not gained much awareness, as the companies with significant environmental and social impacts are rather producing stand alone sustainability reports. The question of which sustainability issues, effects and indicators have significant impact on the financial performance and should thus also be included in the financial report remains open.

1.6.2 The UN System of Integrated Environmental and Economic Accounting (SEEA) and Classification of Environmental Protection Expenditure (CEPA)

Environmental accounting has a micro as well as a macro level; companies are assessing data for internal use as well as for external disclosure. Statistical and environmental protection agencies are collecting this information, aggregating it and providing it for science and environmental politics.

Environmental-economic accounting brings together economic and environmental information in a common framework to measure the contribution of the environment to the economy and the impact of the economy on the environment. The System of Environmental-Economic Accounting (SEEA) is a satellite system of the 1993 System of National Accounts (UN SNA, 1993), a conceptual framework published jointly by the United Nations, the Commission of the European Communities, the International Monetary Fund, the Organization for Economic Co-operation and Development and the World Bank. It consists of an integrated set of macroeconomic accounts and tables based on internationally agreed concepts, definitions, and classifications and accounting rules.

The SEEA 2003 handbook provides a common framework for economic and environmental information, permitting a consistent analysis of the contribution of the environment to the economy and of the impact of the economy on the environment. It is intended to meet the needs of policy makers by providing indicators and descriptive statistics to monitor the interaction between the economy and the environment as well as serving as a tool for strategic planning and policy analysis to identify more sustainable development paths (SEEA, 2003).

Four categories of accounts run through the SEEA handbook. These are

1. Physical and hybrid flow accounts of material and energy (related with material flow accounting on a corporate level). Hybrid accounts link the physical accounts with economic (monetary) flows (called NAMEA matrix).
2. Accounts that portray the environmental transactions in the existing System of National Accounts (SNA) in more detail, e.g. expenditures made by businesses, governments and households to protect the environment.
3. Environmental asset accounts in physical and monetary terms (natural capital in three categories: natural resource stocks, land and ecosystems).
4. Accounts that show how existing SNA aggregates can be modified to account for depletion and degradation of the environment and for environmental defensive expenditure. Such adjustments relate to depletion, so-called defensive expenditures and to degradation.

Material flow accounts (MFA) on a national level are compilations of the overall material inputs into national economies, the changes of material stock within the economic system, and the material outputs to other economies or to the environment. The tradition of economy-wide material flow accounting and analysis goes back to the 1970s (Kneeseet al., 1970). The increasing policy interest in issues of sustainable resource use in the 1990s has resulted in a wider application of economy-wide MFA (see the respective programs and initiatives in the EU, OECD, UNEP, G8, Japan, and China).

The fundamental concept of MFA in SEEA is different to the Input-Output structure on a micro level. SEEA deals with products, natural resources, ecosystem inputs and residuals. The concept of products is taken over from the system of national accounts (SNA). The accounting system of the SNA measures the flows of products (economic goods and services) and shows how in a closed economy some are used to produce other goods and services in the current period (intermediate consumption) or in future (capital formation) and some are used to satisfy current human wants (final consumption). This closed economy must be opened to take account of transactions with the economies of other countries via imports and exports.

Four different types of flows are distinguished in the SEEA (SEEA, 2003, p. 30): ***Products*** are goods and services produced within the economic sphere and used within it, including flows of goods and services between the national economy and the rest of the world. ***Natural resources*** cover mineral and energy resources, water and biological resources. ***Ecosystem inputs*** cover the water and other natural inputs (e.g., nutrients, carbon dioxide) required by plants and animals for growth, and the oxygen necessary for combustion. ***Residuals*** are the incidental and undesired outputs from the economy which generally have no economic value and may be recycled, stored within the economy or (more usually at present) discharged into the environment. Residuals is the single word used to cover solid, liquid and gaseous wastes. Physical flow accounts consist of merging accounts for products, natural resources, ecosystem inputs and residuals, each account being expressed in terms of supply to the economy and use by the economy (Table 1.5).

SEEAs focus is to look at the flow of entities into the economy from the environment and those flowing from the economy to the environment. The environmental inputs flowing to the economy from the environment are divided into natural resources (typically mineral and biological resources) and ecosystem inputs (the water and air necessary for all life forms). The flows from the economy to the environment consist of gaseous, liquid and solid wastes. The term "residual" is used to encompass all these outflows from the economy which use environmental media as a disposal sink and is identical to the terms "waste and emissions" and "non product output" used in EMA.

But, SEEA doesn't make a clear **distinction between materials and products**. It sometimes refers to raw materials only, it sometimes uses the terms materials and products as identical and it doesn't give guidance on the recording of operating materials. While residuals may be clearly identified as non product output, all the materials input side remains vague and inconsistent with accounting terminology and records.

Table 1.5 Physical flow accounts according to SEEA

Inputs	Outputs
Products	Products
Natural resources	Residuals
Ecosystem inputs	

On the **monetary side**, UN SEEA has adopted the Classification of Environmental Protection Expenditures (CEPA) developed by the European Commission and Eurostat (European Commission 2003). The classification includes expenditures whose primary purpose is environmental protection—similar to the information covered under IFAC Cost Categories (3) Waste and Emission Treatment; (4) Prevention and Other Environmental Management; and (5) Research and Development.

CEPA does not cover information contained in the IFAC Cost Categories (1) Materials Costs of Product; (2) Materials Costs of NPO; and (6) Less Tangible Costs and only to a very small degree allows for the information contained in the IFAC Cost Categories (4) Pollution Prevention and Environmental Management as well as (5) Research and Development. Therefore, the information collected under CEPA currently does not include all the information needed for internal management decision making under EMA.

Under CEPA, cost data are first reported by environmental domain (air and climate, wastewater, waste, etc.) and then broken down to distinguish between treatment, prevention and other activities.

Regarding environmental expenditure, the IFAC and UN DSD approach primarily distinguish between treatment and prevention expenditure. It is emphasized that with more sophisticated environmental protection approaches corporations are shifting their emphasis from treatment to prevention and that this shift should be the focus of environmental management and reporting as well.

The CEPA classification in contrast focuses on treatment activities and the impact on environmental media and excludes all activities which make sense to corporations as they pay off. Activities are only to be recorded if the primary purpose is environmental protection and if the expenses don't have a positive return on investment. By this definition most activities that companies are taking for integrated pollution prevention are excluded!

The distinction between End-of-Pipe Treatment and Integrated Prevention is a major achievement in Cleaner Production and highlights the shift in paradigm from emission permits and aftercare to the precautionary principle. Prevention is better than cure is a common saying. The shift in total environmental costs from treatment to prevention started with the widespread application of environmental management systems about 15 years ago, but till nowadays is not adequately reflected in environmental statistics and resultantly in corporate accounting.

In order to understand the SEEA approach to environmental expenditure it is necessary to understand the underlying concept of the "environmental domain of interest" (SEEA, 2003, p. 169): "The two main purposes designated to be of environmental interest are protection of the environment and the management of natural resources and their exploitation. In addition, there are some activities which, though not primarily aimed at protecting the environment, may have environmentally beneficial effects. Damage avoidance and treatment may also be included in the field of interest though these activities are more concerned with rectifying damage already done than with preventing it in the first place. Lastly, and perhaps less obviously, minimization of natural hazards may be included although these are activities to protect the economy from the environment where the others are concerned with protecting the environment from the economy. For simplicity, the expression "environmental activity" is used as shorthand for all the environmentally related purposes just described."

The approach taken by SEEA (p. 170) in identifying environmental activity is to subdivide products and industries into those which are typical, or characteristic, of environmental activity and those which are not. But this neglects the fact that nowadays practically in all sectors environmental management systems have been installed and within them initiatives are being taken to reduce the environmental impact of production and products and in addition develop more sustainable products. It also doesn't solve the problem that products typical of environmental activity may be used for other purposes and some non-typical products may be used for environmental activities.

SEEA tries to solve the issue by introducing a further classification into the supply and use matrix, where the purpose of the expenditure undertaken is identified. This too is subdivided to show the purposes which are environmental in nature, and thus of interest here, and other purposes. In this case the purposes of interest are those listed above:

- Protection of the environment
- Management and exploitation of natural resources
- Environmentally beneficial activities and
- The minimization of natural hazards

But in every day decisions of organizations, investments and current expenditure items are no longer either environmental protection OR production related. It is the success of integrated technologies and management systems (e.g. integrated quality, environment and health and safety systems) that environmental protection is no longer a "satellite system" to general management, but an incorporated strategy and procedure.

Ideally what SEEA (2003) wants to measure are "the expenditures connected with the designated environmental purposes". For practical reasons concerning available data sources, SEEA looks into what has been defined as environmental industries or environmental products (p. 198).

SEEA itself recognizes that "one of the most difficult distinctions to make is whether the primary purpose of the spending is environmental protection, or whether environmental protection is simply a result of decisions taken for some other purpose." It provides the example of spending on equipment which may reduce pollutant emissions but which may also be more energy efficient.

But the solution taken by SEEA is *not to include* the energy efficient equipment, which is not really understandable also from an environmental point of view. This has e.g. let to a strong decline in environmental investments since 1990 (Statistisches Bundesamt, 2006) which is not at all related to a degradation in the state of environment, as companies at the same time have invested in integrated pollution prevention techniques and management systems and actually improved environmental performance in relation to production.

The SEEA approach to environmental expenditure explicitly only "concentrates on steps taken to deal with residuals and does not consider explicitly protection of the environment through means of water and energy conservation or the effects of recycling" (p. 215). In effect, this means that the SEEA approach only focuses on the output of waste and emissions and neglects all activities to reduce the inputs of materials, water and energy. It is thus in complete contrast to the approach of cleaner production and pollution prevention.

The CEPA Definition (SEEA, p. 559) states: "Protection of ambient air and climate comprises measures and activities aimed at the reduction of emissions into the ambient air or ambient concentrations of air pollutants as well as to measures and activities aimed at the control of emissions of greenhouse gases and gases that adversely affect the stratospheric ozone layer. **Excluded are measures undertaken for cost saving reasons (e.g. energy saving)**."

CEPA is designed to classify transactions and activities **whose primary purpose** is environmental protection. The **management of natural resources** (for example, water supply) and the prevention of natural hazards (landslides, floods, etc.) are **not included** in CEPA.

According to SEEA (p. 200) Environmental protection activities are only those where "the *primary purpose* is the protection of the environment; that is, the avoidance of the negative effects on the environment caused by economic activities. Examples include spending by companies on end-of-pipe equipment to reduce or eliminate emissions or make them less hazardous and spending on environmentally protective technology to minimize emissions and pollutant discharges during the production process."

The decision of SEEA to exclude all activities of environmental protection which pay off has in addition contributed to the expectation that environmental protection is costly. But, as environmental prevention projects in the last 20 years have shown very successfully, it is neglected environmental protection and resource management that is costly!

An additional charm of integrated measures is that they pay off for the organization. To exclude them from environmental statistics really only captures a very tiny and the least important picture of pollution prevention! But, companies need to record the costs for resource flows in order to be able to measure this. IFAC therefore explicitly introduced the costs for non-product output. But unfortunately, costs for resource management are still excluded from the environmental expenditure definition of SEEA.

SEEA is currently under revision and better harmonization with the EMA approach to environmental costs is on the agenda. The issues discussed above are also considered in the revision process.

1.6.3 The Guidelines of the Global Reporting Initiative (GRI)

Although EMA focuses primarily on internal management decision making, physical accounting information also is often reported to external stakeholders. Many companies in especially in Europe and Japan include both physical and monetary EMA information in their environmental and sustainability reports.

In the European Union, may companies who have implemented environmental management systems according to ISO 14001 and fulfilled the Requirements of the voluntary European Union Regulation on Environmental Management and Audit Systems, EMAS, and published an externally verified environmental statement have gradually enlarged their disclosure to sustainability reporting, following the guidelines developed by the Global Reporting Initiative, GRI.

The Global Reporting Initiative (GRI) is a large multi-stakeholder network of thousands of experts, in dozens of countries worldwide, who participate in GRI's working groups and governance bodies, use the GRI Guidelines to report, access information in GRI-based reports, or contribute to develop the Reporting Framework in other ways—both formally and informally. GRI has pioneered the development of the world's most widely used sustainability reporting framework and is committed to its continuous improvement and application worldwide. This framework sets out the principles and indicators that organizations can use to measure and report their economic, environmental, and social performance.

The cornerstone of the framework is the Sustainability Reporting Guidelines. The third version of the Guidelines—known as the G3 Guidelines—was published in 2006, and is a free public good. Other components of the framework include Sector Supplements (unique indicators for industry sectors) and Protocols (detailed reporting guidance) and National Annexes (unique country-level information).

Sustainability reports based on the GRI framework can be used to benchmark organizational performance with respect to laws, norms, codes, performance standards and voluntary initiatives; demonstrate organizational commitment to sustainable development; and compare organizational performance over time.

To date, more than 1,500 companies, including many of the worlds leading brands, have declared their voluntary adoption of the Guidelines worldwide. Consequently the G3 Guidelines have become the ***de facto*** global standard for reporting. Companies with more than 300 employees will now have to produce GRI based sustainability reports, the city of Buenos Aires has announced in February 2008. This development follows in the footsteps of the recent announcements by both the Swedish and Chinese governments, who have both mandated sustainability reporting for state owned companies.[1]

The GRI is a collaborating center of the United Nations Environment Programme. The GRI guideline contains performance indicators for the following sustainability issues:

- Economic
- Environmental
- Labor practices and decent work
- Human rights
- Society (including product responsibility)

Within the environmental performance indictors, indicator "Environment number 30" of the GRI Guideline 2006 (GRI, 2006) directly references the IFAC Guidance

---

[1] http://www.globalreporting.org/NewsEventsPress/LatestNews/2008/NewsFeb08BuenosAires.htm

document in its definition but excludes some cost categories: "The compilation of the expenditures in this Indicator should exclude the following categories as defined in the IFAC 'International Guidance Document on Environmental Management Accounting' document:

- Costs of non-product output and
- Fines for non-compliance with environmental regulation

The definition of environmental protection expenditure of GRI states: "Environmental protection expenditure includes all expenditures on environmental protection by the reporting organization, or on its behalf, to prevent, reduce, control, and document environmental aspects, impacts, and hazards. It also includes disposal, treatment, sanitation, and clean-up expenditure." (GRI, 2006).

The costs for NPO are excluded as they are (not yet) commonly recorded by companies and there is no disclosure requirement related with them. Fines are excluded in indicator 30, as indicator "EN 28" explicitly asks for: the monetary value of significant fines and total number of non-monetary sanctions for non-compliance with environmental laws and regulations.

As GRI (2006) is based on IFAC, the environmental expenditure categories also draw the clear distinction between emission treatment and pollution prevention. GRI request the reporting of total environmental protection expenditures broken down by:

- Waste disposal, emissions treatment, and remediation costs and
- Prevention and environmental management costs

Emission treatment according to GRI requires the identification of waste disposal, emissions treatment, and remediation costs based on expenditures related to the following items:

1. Treatment and disposal of waste
2. Treatment of emissions (e.g., expenditures for filters, agents)
3. Expenditures for the purchase and use of emissions certificates
4. Depreciation of related equipment, maintenance, and operating material and services, and related personnel costs
5. Insurance for environmental liability and
6. Clean-up costs, including costs for remediation of spills as reported in indicator EN23 (Total number and volume of significant spills)

Pollution prevention according to GRI requires the identification of prevention and environmental management costs based on expenditures related to the following items:

- Personnel employed for education and training
- External services for environmental management
- External certification of management systems
- Personnel for general environmental management activities
- Research and development

- Extra expenditures to install cleaner technologies (e.g., additional cost beyond standard technologies)
- Extra expenditures on green purchases and
- Other environmental management costs

1.7 EMA Uses and Benefits

Environmental management accounting represents a combined approach which provides for the transition of data from financial accounting and cost accounting to increase material efficiency, reduce environmental impacts and risks and reduce costs of environmental protection. The main areas of application of EMA are internal calculations and decision making.

EMA is particularly valuable for internal environmental management initiatives, such as waste monitoring, cleaner production, supply chain management, eco-design and environmental management systems. As well, EMA-based information is increasingly being used for external reporting purposes. Thus, EMA is not merely one environmental management tool among many. Rather, EMA is a broad set of principles and approaches that provides the data essential to the success of many other environmental management activities. And, since the range of decisions affected by environmental and material flow issues is increasing, EMA is becoming more important, not only for environmental management decisions, but for all types of management activities.

Application fields for the use of EMA data are:

- Assessment of annual environmental costs/expenditure
- Definition of quantified targets for improved environmental performance
- Product pricing
- Budgeting and corporate controlling
- Investment appraisal, calculating investment options
- Calculating costs, savings and benefits of environmental projects and projects to increase material and energy efficiency
- Design and implementation of environmental management systems
- Environmental performance evaluation, indicators and benchmarking
- Cleaner production, pollution prevention, supply chain management and design for environment projects
- External disclosure of environmental expenditures, investments and liabilities
- External environmental or sustainability reporting
- Monitoring and reporting of greenhouse gas emissions
- Other reporting of environmental data to statistical agencies and local authorities

It is however often external pressure that is forcing organizations to look for creative and cost-efficient ways to manage and minimize environmental impacts.

Prominent examples of environmental pressure relevant at the international level include:

- Supply chain pressures, such as large companies requiring their suppliers to comply with the Environmental Management System (EMS) standard of the International Standardization Organization ISO 14001
- Disclosure pressures from various stakeholders for companies to publicly report their environmental performance in annual financial reports or in voluntary environmental or sustainability performance reports, for example, via the guidelines of the Global Reporting Initiative
- Financing pressures via the worldwide growth of climate, ethical and socially responsible investment (SRI) funds, investment rating systems such as the Dow Jones Sustainability Index and investment policy disclosure requirements
- Regulatory control pressures, for example, the REACH Directive, a European Union (EU) regulation that requires monitoring of chemical substances and products (European Comission 2007)
- Environmental tax pressures, for example, various government-imposed environment-related taxes such as carbon taxes, energy use taxes, landfill fees and other emissions fees
- Cap and trade pressures, such as the emissions cap and trading aspects of the Kyoto Protocol (1997)

Improved and harmonized data quality is essential for corporations as well as for aggregated statistical analysis, as they provide the ground for several decisions, from investment choices to scientific projects and political instruments and allow better benchmarking. In addition, the time needed for data assessments and aggregations can be reduced significantly, as well for corporations as for statistical agencies.

EMA data can be collected, analyzed and used at different system boundaries, such as:

- The entire organization
- A particular business group
- A single site or facility
- A particular product or product line
- A specific cost center
- A particular process or equipment line
- A particular raw or operating material
- A specific waste stream

From an accountant's point of view, the most likely starting point for EMA is the list of accounts, which is the most common source of cost information in all organizations. Working with the list of accounts allows an assessment of site-wide or organization-wide annual costs related to environmental issues. This assessment alone will probably lead to improvements in the accounting, information and control systems, as it will soon highlight problems such as inconsistencies in the posting to accounts, missing information or in assumed scrap percentages. From an accountant's point of view the top down approach is the most appropriate to ensure completeness and consistency of information systems.

From an environmental manager's point of view, the desired starting point may be the analysis of a particular waste stream. A production manager might be the most interested in monitoring a particular product line or set of production equipment. These more detailed analyses will require going deeper into the accounting systems—looking at cost center reports, calculations of production costs and product prices, statistics on scrap and returned poor quality product, recipes from the production planning system, inventory reports, waste reports, as well as energy, water and materials balances. It might even require installing actual measurement facilities for missing information.

But the starting point doesn't really matter. The most important task is to make sure that ALL relevant and significant costs are considered when making business decisions. In other words, environmental costs are just a subset of the bigger cost universe that is necessary for good decision making. Environmental costs are part of an integrated system of material and money flows throughout a corporation, and not a separate type of cost altogether. Doing environmental management accounting is simply doing better, more comprehensive management accounting, while wearing an environmental hat that opens the eyes for hidden costs. Therefore, the focus of material flow accounting is no longer assessing the total environmental costs, but on a revised calculation of production costs on the basis of material flows.

Chapter 2
The Input Side of the Material Flow Balance

Chapter 2 describes the input side of the material flow balance. The physical accounting information collected under EMA is a prerequisite for the calculation of many environment-related costs. Mass balances in volumes, energy content and liters and materials flow accounting in monetary terms are the basis for EMA assessments. Inputs are any energy, water or other materials that enter an organization. Materials Inputs comprise raw and auxiliary materials, packaging materials, merchandise, operating materials, water and energy.

2.1 Overview on Material Flow Balances

In the financial balance sheet assets and liabilities balance off to zero. Likewise the physical material flow balance of manufacturing companies in theory must balance off to zero as well. In practice, very few organizations are able to calculate their material flows to this degree.

But, whatever has left a company not as a product is a sign of inefficient production and must by definition be waste and emissions. Determining the material flows for, at least, raw and auxiliary materials is therefore imperative for an environmental as well as production oriented cost assessment. The concept of material flow balances has been developed for the production sector and is less applicable to the service industry and the agricultural, forest and mining sector (Environmental Protection Agency, 1999, Jasch, 2002).

The material purchase cost of wasted materials comprise the most important environmental cost category, accounting for 40–70% of total environmental costs, depending on the value of raw and operating materials and the labor intensity of the sector. Cost savings are often feasible in the material costs category, but the material flows have to be made transparent and traceable therefore. Companies put a lot of effort into exploring saving potentials by reducing the number and costs of employees, but are rather advised to spend more time tracing the losses of materials purchased.

C. Jasch, *Environmental and Material Flow Cost Accounting*: Principles and Procedures, Eco-Efficiency in Industry and Science 25,

Before waste and emissions occur, the materials concerned have been

- Purchased (materials purchase costs)
- Transported, handled and stocked (costs for stock management, handling and transport)
- Processed in various production steps (equipment depreciation, work time, auxiliary and operating materials, costs for finance etc.)
- Collected as scrap, waste, etc., sorted, transported, treated, transported, stocked, again transported (personnel, external services and fees) and finally
- Disposed off (disposal fees)

Corporations thus pay three times for non-product output at purchase during production and at disposal.

Improvement of environmental performance is based on the evaluation of material flows through an input-output analysis of the material flows in **physical terms**. The system boundaries for the first assessment may be the organization, as many invoices and data are recorded only on this level and later can be further divided into sites, cost centers, processes, and products.

Table 2.1 shows the structure of the material flow balance. First the total consumption of raw-, auxiliary-, and operating materials by material groups are recorded from the list of accounts and cross checked with additional records, e.g. from stock management. Then the related volumes are added to the input side, to the degree available in the current system. The product and non-product output needs to be recorded only in physical terms, as the monetary evaluation is done in the monetary part of the EMA assessment. But for the Input side it makes sense to collect both

Table 2.1 Structure of the material flow balance

Input in physical and monetary terms	Output in physical terms
Raw materials	Product
Auxiliary materials	Core product
Packaging	By-product
Operating materials	Waste
Merchandise	Commercial waste
Energy	Waste for recycling
Gas	Hazardous waste
Coal	**Wastewater**
Heating oil	Amount in m^3
Gasoline	Heavy metals
District heating	CSB
Renewable resources (Biomass, Wood)	BSB_5
Solar, Wind	**Air emissions**
External produced electricity	CO_2
Internally produced electricity	CO
Water	NO_x
City water	SO_2
Well water	Dust
Spring water	FCKWs, NH_4
Rain/surface water	Ozone destroying substances

physical and monetary values at the same time in order to insure consistency of the data. A problem often encountered is that the monetary system records the data including inventory changes while the physical system records the actual flow and differs from the monetary recording without easy options to reveal why.

The material flow balance is based on the idea that what goes into an organization must (at some point) come out. It includes all the inputted materials, as well as the resulting amounts of products and NPO. The purchased input is compared to the production volume, the sales statistics, as well as the records of waste and emissions. The goal is to improve the efficiency of material use, what leads to both economic and environmental improvements.

Materials Inputs are any material, energy or water that enters an organization. Definitions of the various Materials Input categories are given below. For materials recorded in stock management, not the values for materials **purchased**, but **consumed for production** should be used respectively for both the physical and the monetary values. Chapter 7 provides more information on the linkages between the physical and monetary information systems.

In manufacturing companies, most Materials Inputs are eventually incorporated into physical products (including by-products and packaging). These have potential environmental impacts when they leave the manufacturer, for example, if a product leaches toxic materials after it has been disposed of in a landfill at the end of its useful life. In addition, the extraction of all natural resources has environmental impacts, such as ecosystem disturbance at the extraction site. Thus, the overall materials-related environmental impacts of a manufacturer's product during its life-cycle from materials extraction, several manufacturing steps, use at the customer and final disposal may often outweigh the environmental impacts during production.

The purchase costs of Materials Inputs that are converted into products, by-products and packaging are directly assessed when providing the mass balance as the sources of information are probably identical and should be consistent. The physical accounting side of EMA provides the information on the amounts and flows of materials and products needed to assess such costs. Once the material inputs, flows and costs have been assessed on a company level, they can be further separated on a cost center or material specific level.

These cost data help an organization to cost-effectively manage the materials-related environmental impacts of its products. For example, it might consider replacing a toxic product ingredient with a less-toxic, cost-effective alternative. This data is also essential for investment appraisal.

2.2 Raw Materials

Raw (and Auxiliary Materials) are Materials Inputs that become part of an organization's final physical product or by-product. Raw Materials are the major product components (for example, the wood used in furniture manufacturing). In many companies, warehouse management and production planning systems monitor their purchase and input into production.

In most companies, raw materials are already being recorded in a very detailed manner via accounts as well as material stock numbers, warehousing, production planning systems and cost accounting. Thus, material purchase costs and quantities consumed are often available. If needed, average prices can be used to calculate the weight values. The assignment of material stock numbers to financial accounts is sometimes not treated systematically and should be clearly defined. Raw materials and auxiliary materials are often assigned to separate accounts, since they usually contain rather homogenous substances and significant purchase values.

2.3 Auxiliary Materials

Auxiliary materials become part of the products, but they are not considered its main components (e.g. glue in a table or shoe, salt in a cake). As they become a product component, most of their input should be on the product, but a loss percentage has to be estimated, if no measurements are available.

Many organizations don't clearly distinguish between auxiliary and operating materials, but record them on joint accounts and don't monitor the actually use and losses in production via production planning systems and technical monitoring systems. Technologies, which significantly increase the efficiency of production, can therefore not be adequately assessed by investment appraisal technologies.

The materials input of auxiliary materials should be recorded separately for each material group and loss percentages calculated or estimated. The employees at the related production lines often can provide good estimates, which are unknown to the financial departments. Eventually these materials should be included into inventory managed warehousing and process monitoring.

2.4 Merchandise

Several organizations purchase products for trade with little or no additional processing. Products parts are produced at external suppliers and just added to the final product without processing. It can therefore be assumed, that little waste is related with merchandise (besides packaging). If this is the case, merchandise only needs to be recorded, if a consistent mass balance in volume is attempted, as merchandise can constitute a major part of the products sold.

In other business sectors, merchandise can be related to significant environmental impact and costs, as it needs special handling and storage (e.g. cooling of food) and may be required to be disposed of as it has outlived its useful shelf life. In this case, handing of merchandise may require a cost monitoring on its own in order to be able to collect the associated costs and the amounts on the product and in waste should be recorded or estimated.

2.5 Packaging

Packaging materials show up on the input and the output side of the material flow balance.

Packaging materials are purchased for shipping organizations final products. In several European countries with licensed packaging systems, these volumes are recorded in detail by material groups and included in production planning systems, warehouse inventory management and even external reporting to licensing agencies.

However, some EMA case studies have shown that consistency of recording for packaging materials can be improved, as sometimes packaging material is recorded with material numbers, but is not recorded in the warehouse inventory. Frequently, some packaging material purchased is not assigned to the corresponding account and material stock numbers, but is subsumed under other operating costs or under overhead. Some organizations don't have clear rules for which material numbers should be posted on what account and resultantly, the total material input of packaging materials can't be traced from the accounting records.

Packaging purchased for an organizations product will mostly leave the organization together with the product, but again, a certain loss percentage (e.g. due to repackaging for specific destinations) needs to be estimated if no records are available. For multiple use packaging systems (such as pallets) the annual purchase of additional equipment can be used as estimate as well for the material input as for the waste output.

Packaging material delivered by suppliers together with raw-, auxiliary and operating materials is included in the purchase price and while often generating costs a second time via disposal costs, is only rarely recorded separately, despite the fact that it constitutes a large share of waste incurred. While product packaging leaves the company together with the product and must still be disposed of by either the retailer or the consumer, the company must dispose of supplied packaging material unless it is shipped back to the supplier.

The material flow balance thus contains

- Purchased packaging for an organizations products on the input side
- Products including packaging on the output side and the
- Loss of products packaging as well as the
- Packaging for raw, auxiliary and operating materials under non-product outputs (typically as solid waste)

2.6 Operating Materials

Operating Materials are Materials Inputs that an organization purchases and uses but that do not become part of any physical product delivered to a customer. Organizations in the production sector use operating materials like chemical catalysts, cleaning materials, greases, industrial gases, glues, paint, maintenance materials, small tools, etc. For organizations in the service sector all Materials Inputs must be Operating Materials, for example, fuel for transport services.

As Operating Materials do not become part of any physical product, they by definition become Non-Product Output (Waste and Emissions) when they leave the organization.

They may contain harmful and toxic substances, e.g. for use in laboratory or workshop, which often have to be disposed of separately as hazardous waste. In many companies they are not traced by the storage administration system but are recorded as expenditures at purchase. Very rarely is their consumption assigned to a cost center, which makes subsequent tracing difficult. While their consumption is recorded in the production overhead cost surcharges, a comparison with actual consumption is rarely done. If these materials are not included in the material stock management system, for the first round of an EMA setup it is recommended to simply record their total purchase value and not try to estimate their volumes, but consequently install a recording procedure for the future.

The distinction between Operating and Auxiliary Materials is vital, as the strategies for waste prevention are different. While for Auxiliary Materials the target is to reduce the loss percentage, for the Operating Materials the target can only be to use as little of them as efficient as possible and to choose the ones with the least environmental impact.

Other Operating Materials, which are often not regarded when installing an EMA system in the first round, are office supplies, building cleaning supplies etc. as the focus should be on production and the most significant costs and environmental impacts.

When recording operating materials, it should be ensured that no services and labor costs are entered into the accounts. These should be recorded separately. In principle, all profit and loss accounts need to be examined for material flows in order to compile a complete material flow balance sheet. In practice, for the start up, available data from material accounting and the technical departments as well as estimates will determine the scope. The assessment should result in improvement options for data recording and thus gradually improve the availability of material flow information.

Operating materials constitute great potentials for saving as they have often been neglected before. Not many companies already record oils, lubricants, chemicals, paints, varnishes, diluting agents, glues, cleaning agents and other operating materials via material numbers and warehouse inventories. In most cases, there are no separate accounts for operating materials and they are not accounted for in production lists or production planning systems.

Cost center assignment can also be improved in many ways. In most organizations, the consumption of operating materials is not recorded on production cost centers so it is practically impossible to trace who has used how much of them. In cost calculation, only estimates are used for the calculation of product prices, but hardly ever somebody checks if these estimates confirm to real consumption.

Frequently, operating materials disappear into overhead and cannot be traced in detail. It is therefore advisable to record and classify, to the extent possible, via material numbers or posting to separate accounts, at least those operating materials which are related to hazardous waste disposal or other waste flows. Large quantities of many of these substances "disappear" into accounts like "other operating costs". It is thus very difficult to trace their consumption without having to go back to original invoices or keeping separate notes.

Repair materials and spare parts, as well as maintenance, are often recorded under entirely different categories. Since the repair and maintenance shop as well as the laboratory are particularly critical parts of the company with regards to environmentally relevant substances and the production of hazardous waste, it would be desirable to ensure that the materials used are disposed of in an appropriate manner and that, without exception, they are recorded in the inventory. The materials use can be kept on file through the special cost centers for shop and laboratory.

A similar approach applies to cleaning agents, which not all firms record on separate accounts. Ecological relevance and quantities will determine the degree of detail of those records.

As a cross check of the amounts recorded as non product output, the material content of waste can be assessed and recalculated to the input materials. For solid waste, the material input is comparatively easy to assess. But some of the purchased materials do not end up in disposal, but are converted into air emissions or can be found in waste water.

2.7 Energy

The Energy category includes energy of all types that an organization uses: electricity, gas, coal, fuel oil, district heating and cooling, biomass, solar, wind and water. For some utilities Energy may constitute a product but, in general, Energy is viewed as an Operating Material, in that the Energy is not intended to become part of a physical product but is instead used for running equipment, etc.

Energy purchase can easily be traced via the respective invoices and is often monitored already. Energy consumption is relevant to all businesses and is important for the calculation of various air emissions. Energy input should be quoted consistently in kilowatt hour.

The energy purchased should be adjusted by recording energy sold to others (e.g. electricity, steam) as a product output. Internal production is not considered on the system boundary of the company fence and the profit and loss account. The energy balance is calculated separately from the mass balance.

2.8 Water

Water input consists of all water from all sources, such as rainwater, groundwater, surface water from rivers and lakes, regardless of how the water is obtained (for example, private wells or the public water supply system). Water used for cooling purposes should be recorded separately. Water input can be obtained from water supply invoices and must be estimated for supplies from own wells and surface water unless monitoring systems are installed.

In some manufacturing sectors, such as food processing, water may be part of the final physical product (much like Raw and Auxiliary Materials), while other water is never intended to go into a final product but is used for other purposes, such as cooling or cleaning (much like Operating Materials). Thus, some water may leave a manufacturing organization in the form of physical product, but mostly it will leave as Waste Water or Air Emission.

Chapter 3
The Output Side of the Material Flow Balance

Chapter 3 describes the output side of the material flow balance, which is assessed only in physical, not monetary terms, as the related costs are traced separately. Outputs are all products, wastes and emissions that leave an organization. Product Outputs are products and by-products including their packaging. Non-Product Outputs comprise solid waste, wastewater and air emissions. Any Output that is not a Product Output is by definition a Non-Product Output (NPO) and comprises waste and emissions in solid, liquid and gaseous form.

3.1 Products and By-products

This category is relevant only for organizations that produce a physical product, such as resource extractors or manufacturing operations.

Products include all physical products and their packaging. By-products are products incidentally produced during the manufacture of the primary product. In many organizations the boundaries between products, by-products and waste are not well defined, and depend partially on how well an organization separates by-products and waste. Whatever is being sold and shows up as earning in the accounts can be considered a by-product.

The quantity of products produced in a fiscal year can usually be determined from productions statistics and final stock records; however, sometimes it has to be calculated from turnover. It is important to note that turnover is only a part of total production Once a product has been manufactured, there will be losses during warehousing, the quality department may discard some production and the company may consume a certain amount itself. But ideally, all losses occurring between production and turnover should be assigned to non-product output.

In some business sectors a recording of volumes of products produced is not installed (e.g. turnover measures in pieces of cars but not in their volume). In such cases, material flow accounting lacks the basis for the calculation of a mass balance and for a consistency check of scrap figures and loss estimates. At least for a pilot

C. Jasch, *Environmental and Material Flow Cost Accounting*: Principles and Procedures, Eco-Efficiency in Industry and Science 25,

project time frame, a complete listing and recording of the most significant material flows in volume is recommended.

3.2 Non-product Outputs (Waste and Emissions)

Any Output that is not a Product Output is by definition a Non-Product Output (NPO) in the form of solid waste, wastewater or air emissions. Waste and Emissions are generated by raw and auxiliary materials as well as by operating materials including energy and water.

Raw and auxiliary materials, packaging and merchandise, that were intended to become products, to a certain degree become waste and emissions. The reasons are production inefficiencies, scrap, poor maintenance, inefficient operating practices, production losses, product spoilage, poor product design, quality deficiencies or other reasons. For all these, loss (scrap) percentages should be measured, calculated or estimated.

Operating materials are by definition not part of the product and therefore must become NPO and end up in waste and emissions.

3.2.1 Waste

Solid Waste can be distinguished into materials for recycling, such as waste paper, plastic, and glass and scrap metal (which might classify as by-products, if they are being sold), municipal waste and hazardous waste.

Hazardous Waste is often defined by national law and has to be disposed of separately as it contains infectious, flammable, toxic, carcinogenic and other harmful substances. It can occur in solid form (such as discarded batteries), liquid form (such as waste paint and solvents) or mixed form (such as wastewater treatment sludge).

In the first round of material flow cost accounting it is unlikely, that the mass balance will be equal as complete records for waste and emissions (and the volumes of material inputs) will probably not be available. However, based on a two weeks detailed assessment and measurement, estimates of the annual quantities by type of waste should be calculated.

Once the types of waste generated and their origins (which production processes are responsible for waste generation) have been determined, options can be developed to prevent or recycle waste. In many case studies separation of waste at the source of origin has resulted in possibilities for reuse within the organization or options of selling former waste as by-product. This saves money and reduces environmental impact.

Treatment of waste in the **financial accounting** system is another issue. Its needs no mentioning, that expenditure for waste handling and disposal and revenues from selling metal scrap etc. should be posted on different accounts. Additional records will be needed to have complete records of the costs of waste management by each waste fraction. In most companies the environmental manager is in charge for these records. But experience has shown that often he is only aware of the direct disposal fees and has no information on additional handling costs which may disappear on the several supplier accounts for external services. The accounts for cleaning, transportation, maintenance and third-party services may need examination whether they contain invoices that should be assigned to disposal costs.

In addition in some countries and industry sectors specific waste (e.g. scrap wood for burning) is given free of charge to employees and local residents. Again, records for the quantities should be kept.

Dealing with waste in **cost accounting** is also worth consideration. In most companies disposal cost are not assigned to cost centers and disappear in general overhead. Several companies have decided to install a cost center for environmental management and post them there, as the environmental manager needs that information at the most. But is he really responsible for the waste generated? In some organizations systems have been installed where the cleaning department keeps records on the amounts and types of waste collected from the different production processes or even the separate production shifts. The amounts and costs of waste are then levied back to the different production cost centers just like any other direct cost. Resultantly the cost calculation for each product specifically reflects the costs of disposal attached to it. At the same time the different production shifts have an interest in waste minimization.

The purchase department also plays a key role in waste disposal. Combining procurement and disposal responsibilities changes awareness of the purchase department and has often led to the application of multi-use packaging systems and take-back obligations negotiated with suppliers. A measuring and weighing system should be installed at Purchasing and Delivery, in order to obtain information about the quantities of waste disposed of. In one of the case studies in Austria the first investment initiated by the EMA project was a scale at the Incoming Store. All waste should be calculated or converted into metric tons.

Companies that operate their own disposal or incineration plants should also keep records on the amounts and types of waste processed.

3.2.2 Waste Water

Waste water is all water that exits an organization apart from water contained in its products. Waste water streams contain municipal waste water, direct streams into rivers or the sea as well as surface water.

The amount and content of wastewater is often not monitored on a regular basis, as such monitoring is only required for specific sectors and specific waste water streams. Many countries require spot checks for companies that pass waste water directly into rivers or the sea, from which annual quantities of contaminants of some kind, such as high biological oxygen demand (BOD), total suspended solids (TSS), nutrients (such as phosphates), excess heat and toxic materials (such as solvents, pesticides or heavy metals) can be estimated.

The volumes of water input and output should not be considered in the mass balance but calculated separately in a water balance, if water flows are significant for the specific business sector.

3.2.3 Air Emissions

Air Emissions are air streams contaminated with problematic levels of pollutants. Examples include emissions of energy combustion, such as nitrogen oxides, sulphur dioxide, carbon monoxide, particulate matter consumed and volatile organic compounds, as well as other pollutants such as metal particulates. Air emissions can also include radiation, noise and heat.

Waste heat and air emissions are typically estimated based on the type of energy, materials and processes used (e.g. solvents, cleaning agents). Applying commonly used conversion factors for fuel emissions is recommended.

The fossil fuel energy (primary energy) used to generate the electricity purchased by an organization depends heavily on the local or national energy mix and technology used to generate electricity. Several countries publish national conversion factors for the corresponding primary energy input. Country specific data for fossil, nuclear- and hydropower electricity generation can be used to calculate the specific primary energy input and related CO_2 and other emissions. As energy markets are being liberalized, the situation will become even more complex.

The Kyoto Protocol (1997) covers industrial and energy linked global warming gas emissions. The main substances are Carbon Dioxide, Methane, Nitrous Oxides, Sulphur Hexafluoride, Perfluorcarbons and Hydrofluorcarbons, resulting from fuel combustion, process reactions and treatment processes. All greenhouse gas emissions should be calculated in metric tons of CO_2 equivalent.

CFC emissions contribute to the depletion of the ozone layer. The Montreal Protocol (1987) covers ozone-depleting substances and standardizes their ozone depletion potential (ODP) in relation to the reference substance CFC-11. Once the volumes purchased have been assessed, conversion factors (see Annex) should be applied to calculate ozone-depleting emissions in metric tons of CFC11 equivalents.

Even if a certain substance is not emitted, it should be recorded in the input–output balance as n.r. (not relevant). This will indicate to the internal or external user that certain substances were not omitted from consideration, but were actually not emitted.

Chapter 4
Environmental Performance Indicators

Chapter 4 deals with environmental performance indicators, which for the operational system are directly derived from the input output material flow balance. The definitions provided in the ISO 14031 standard as well as the related indicators recommended by the Global Reporting Initiative are described. In addition the chapter discusses requirements and system boundaries for indicator systems and specifically addresses the problem of finding meaningful denominators for performance indicators. The chapter concludes with a case study from the brewery in Murau which calculates savings based on their environmental performance indicator system.

Environmental performance indicators condense environmental data into relevant information that allows monitoring, target setting, tracing performance improvements, benchmarking and reporting. Several publications and pilot projects highlight their relevance for environmental management systems, improving material efficiency and flow management, and detecting cost saving potentials and quantifying performance targets.

Environmental performance indicators supply the operational level as well as top management with the information required for decision making. On this basis, well-grounded targets for environmental performance improvement can be identified, quantified and achievement monitored. The process of gathering physical data to be reported is often not called EMA at all, or even called accounting, as the experts on much of this physical flow information tend to be the personnel in purchase, production and environmental department, rather than those in accounting.

The strengths of environmental performance indicators (EPIs) are quantification of risks and trends and benchmarking with previous years and other sites. If monitored regularly they thus serve as an early warning system. The comparison of environmental performance indicators within one company or externally with other companies or competitors, so-called benchmarking, offers options to identify improvement potentials.

C. Jasch, *Environmental and Material Flow Cost Accounting*: Principles and Procedures, Eco-Efficiency in Industry and Science 25,

4.1 ISO 14031—Standard on Environmental Performance Evaluation

The international standardization organization, ISO, published a standard on environmental performance evaluation, ISO 14031 (ISO, 2000), in connection with its standard on Environmental Management Systems, ISO 14001 (ISO, 1996). The standard is based on a material flow balance for the operational system as described in the previous chapters. In addition, it lists indicators for the management system and for the condition of the environment outside the organization.

Environmental performance indicators, **EPI**s are defined as follows in ISO 14031:

> *"**OPI**, Operational performance indicators, provide information about the environmental performance of an organization's operations."*

Operational performance indicators form the basis of evaluation of environmental aspects. They directly relate to the input-output material flow balance. Examples are material, energy and water consumption, waste and emissions in total amounts and in relation to production volumes. OPIs are an important basis for internal and external communication of environmental data, e.g. in environmental statements in accordance with the EU EMAS-Regulation or in internal reports to inform operational staff.

> *"**MPI**, Management performance indicators, provide information about the management's efforts to influence an organization's environmental performance."*

Management performance indicators indirectly measure the environmental protection efforts taken by a company and the results achieved with regard to influencing its environmental aspects. Examples are the number of environmental audits, staff training, supplier audits, cases of non-compliance, certified sites etc. They provide useful information and allow target setting to improve the environmental management system, however they don't relate to the actual external environmental impact or internal environmental aspects. An exclusive use of MPIs for evaluating environmental performance is not recommended in ISO 14031, as they do not reveal the significant environmental impacts and may even camouflage them.

> *"**ECI**, Environmental condition indicators provide information about the local, regional, national or global condition of the environment."*

Environmental condition indicators directly measure the quality of the environment. They are used to assess the impact of air emissions on air or water quality. The environmental conditions around a company, such as water and air quality, are typically monitored by government authorities. Only if one particular company is the sole or main polluter in a region, monitoring by individual companies may be requested by law or may make sense also voluntarily, e.g. noise for airports, air quality for power stations, and water quality for pulp and paper industries. Since the quality of environmental media such as air, water, soil and the impacts of human activities (e.g. over fertilization of water, reduction of biodiversity, greenhouse effect) depend on many factors (emissions of other companies, of power plants, households and traffic), the measurement and recording of ECIs are primarily performed by public institutions.

Global and national indicators for the evaluation of environmental quality are mostly termed "environmental indicators" or "environmental condition indicators" and are not referred to as "performance indicators".

4.2 Environmental Performance Indicators of GRI

The Global Reporting Initiative, GRI, published its latest version of its global reporting requirements in October 2006 (GRI, 2006). They contain a set of sustainability performance indicators which are supplemented by indicator protocols which specify the measuring and disclosure requirements in detail. The GRI definition of environmental expenditure directly references the IFAC EMA guidance document.

The sustainability reporting guideline of GRI (GRI, 2006) lists indicators for

- Economic performance
- Environmental performance
- Social performance
- Human rights and
- Society (including product responsibility)

The indicators are accompanied by detailed indicator protocols which specify their content and ensure comparability with other reporters. The environmental performance indicators (abbreviated with EN in the GRI guideline) that relate to the material flow balance are shown in Table 4.1. Materials as defined in indicator EN 1 refer to all raw, auxiliary, packaging and operating materials and also include semi-finished goods. The IFAC and GRI approach to physical materials accounting is thus compatible.

4.3 General Requirements for Indicator Systems

Decision making at many different levels can be supported by Environmental Performance Indicators (EPIs). EPIs can be created from purely physical information collected under EMA (for example, the total amount of wastewater treated each year) or purely monetary information collected under EMA (for example, the total cost of wastewater treatment each year). Physical EPIs and monetary EPIs can also be combined into cross-cutting EPIs that link the two types of information (such as the wastewater treatment costs per unit customer service each year).

EPIs monitor a company's effectiveness and efficiency of resource management (Jasch & Rauberger 1997). This applies mainly to physical resources like materials,

Table 4.1 GRI environmental performance indicators related to physical materials accounting

	Materials inputs		**Product outputs**
	Materials		No disclosure requirements
EN 1	Materials used by weight or volume		**Non-product outputs (emissions, effluents and waste)**
EN 2	Percentage of materials used that are recycled input materials	EN 16	Total direct and indirect greenhouse gas emissions by weight
	Energy	EN 17	Other relevant indirect greenhouse gas emissions by weight
EN 3	Direct energy consumption by primary energy source	EN 19	Emissions of ozone-depleting substances by weight
EN 4	Indirect energy consumption by primary source	EN 20	NO_x, SO_x and other significant air emissions by type and weight
	Water	EN 21	Total water discharge by quality and destination
EN 8	Total water withdrawal by source	EN 22	Total weight of waste by type and disposal method
EN 10	Percentage and total water volume of water recycled and reused	EN 23	Total number and volume of significant spills
		EN 24	Weight of transported, imported, exported, or treated hazardous waste and percentage of transported waste shipped internationally
		EN 29	Significant environmental impacts of transporting products and other goods and materials used for the organisation's operations, and transporting members of the workforce

but can also be linked to other resources like personnel and money. Indicators are most useful and meaningful if they are

- Monitored over time
- Comprised of two variables, an absolute measure and a reference measure
- Comparable across sites and companies

The process for setting up an indicator system has been described in several projects and publications. The following principles should be applied when installing an indicator system (Jasch & Rauberger 1997).

4.3.1 Relevance

The indicators should adequately reflect the significant environmental aspects and impacts of the organization and be selected by the people in charge of controlling, monitoring and target setting. Data should be collected only, if it is to be used internally or for external disclosure.

4.3.2 Understandability

Indicators must be clear and correspond to the user's information needs. If indicators become too complex, for example aggregating several items by complex mathematical calculations, people lose understanding of their meaning and how the indicator may be influenced. People in charge of activities with environmental impact must understand how an indicator can be influenced.

4.3.3 Target Orientation

The indicators should correspond to environmental improvement targets.

4.3.4 Consistency

Comparable and reliable EPIs throughout an organization can only be achieved by standardization of relevant environmental and financial indicators. The same method must be used to calculate EPIs across a company, defining in detail the database and calculation procedure for each variable. In addition, the method for calculating EPIs should be consistent with the financial information system and indicators.

4.3.5 Comparability

Indicators must allow comparison over time and with other sites and business units. Thus, the calculation principles, data sources and definitions for each nominator and denominator must be defined to make sure that the data-base is consistent across reporting units and time series. For comparison, establishing the same data collection principles in every period, referring to comparable intervals and measuring comparable units, is essential.

4.3.6 Balanced View

An indicator system should measure changes and cover all significant environmental aspects and impacts. For all major categories of the material flow balance indicators should be defined. A common trap is to use only data available and base the indicator system on for example 20 indicators for waste, as it is being monitored, but neglecting air and water emissions and material input, simply because data is not available.

Table 4.2 Environmental performance indicator system

	Absolute quantity	Relative quantity by production output (PO)
Production output (PO)	kg, L	
Raw material input	kg	kg/PO
Auxiliary material input	kg	kg/PO
Packaging material input	kg	kg/PO
Operating material input	kg	kg/PO
Energy input	kWh	kWh/PO
Water input	m^3/L	m^3/PO
Waste	kg	kg/PO
Waste water	m^3/L	m^3/PO
Specific pollution loads	kg	kg/PO
Air emissions	m^3	m^3/PO
Air emissions load	kg	kg/PO
Other denominators		
Number of employees	Number	
Turnover	Monetary value	
EBIT	Monetary value	
Production hours	Time	
Workdays	Days	
Building area	m^2	
Management performance indicators		
Number of achieved objectives and targets		
Number of non compliances or degree of compliance with regulation		
Number of sites with certified environmental management systems (EMS)		
Percent of turnover of production/products with environmental labels		
Percent of turnover from EMS certified sites		

4.3.7 Continuity

Indicators become more meaningful if they are monitored by the same method over longer periods. The time intervals for assessment (daily, weekly, monthly, yearly) should allow timely intervention in case of undesired developments (like break down of automatic sensors for water and material supply) and prevent outdated information. If indicators are calculated too infrequently or at too long intervals, there is little relation to current performance.

As a general outline for generic indicators that can be applied throughout all sectors, the following indicators are recommended. Sector specific, more detailed indicators may be valuable, but aggregation to the general categories should be possible. The indicator system should cover all major input and output categories (Table 4.2).

4.4 System Boundaries for Performance Indicators

Performance indicators can be useful for many system boundaries of the organization, site specific and further down to cost centers and production processes (Jasch, 1988). Each decision maker requires information for his scope of responsibility. Thus, caution must be given to aggregation without double counting and to insure consistency of calculation. It has therefore often been advisable to establish indicator protocols like the GRI Guideline, which in detail define how an indicator should be calculated and the data sources and measuring techniques. Experience form case studies has shown that different business units or sites otherwise tend to different interpretations.

Data on different system boundaries serves different purposes. The most common system boundaries for environmental performance indicators are

- Production processes or product lines
- Cost centers
- Business units
- Production sites
- Companies within a corporation
- Corporate level

Indicators derived from the lower organizational level **departments, processes, cost centers** may be suitable primarily as a monitoring instrument for the respective departments, serving as earl warning system against spills and leakages and for monitoring of scrap. Assessments should be at shorter time intervals, e.g. quarterly, monthly or weekly, in order to determine weak points and to take corrective measures in time. The main inputs of raw and auxiliary materials, and energy as well as the major sources of emissions should be monitored on a process level.

Site and corporate indicators serve as general performance information for management over a longer period of time and for annual reporting purposes. They allow benchmarking of sites and target setting at corporate level.

4.5 The Problem of Finding a Meaningful Denominator

From an environmental point of view, expressing indicators in **absolute terms** is the most meaningful way of recording as the total consumption of resources and the total impact on the environment are made visible (e.g. the consumption of auxiliary materials in kilogram or the quantity of waste water in cubic meter). For comparison however, a relation to production volumes or other significant denominators is necessary. Relative indicators present the environmental performance of an organization in relation to its size, to production output or turnover or to the number of employees. While absolute indicators describe the total environmental burden, **relative indicators** allow monitoring of efficiency improvements. Absolute and relative indicators are two sides of a coin and are both useful. The implications of relative indicators cannot be judged without the absolute data base and vice versa.

It must however be stated, that often the efficiency gains of relative performance indicators are offset by increasing production volumes, so that the total impact on the environment actually increases.

Indicators can be presented in the following ways:

- Absolute figures, like tons of waste per year.
- Relative figures, compared to another parameter. The most common denominators are production volumes, production hours, sales (turnover) and number of employees.
- Percentages or indexed, in relation to a baseline, like hazardous waste as percent of total waste, or hazardous waste as percent of the previous year.
- Aggregated data; of the same type, but from different sources, expressed as a combined value, such as total tons of SO_2 emissions from five production sites, aggregated to the corporate level.
- Weighted, data multiplied by a factor related to its significance, prior to aggregating or averaging.

The **environmental performance indicator matrix** shows possible combinations of absolute environmental performance indicators with relevant denominators to obtain significant relative indicators. The longitudinal axis provides examples of absolute indicators (basic data from the material flow balance), which can be related to the relevant denominators on the horizontal axis. The check mark "□" indicates useful combinations, the actual choice depends on the business sector and company specifics. In addition to the main categories of the input-output mass balance, the matrix also contains environmental management performance indicators. Depending on the production range of a company, other variables may also be useful (Table 4.3).

Table 4.3 Environmental performance indicator matrix

	Production volume	Cost centre	Machine time	Material input	Number of employees	Working days	Shifts/ working hours	Site area (m^2)	Revenue or turnover	Production costs
Material input	√	√	√				√		√	√
Packaging	√	√	√	√						√
Energy input	√	√	√			√				
Water input	√		√		√	√				
Detergents	√	√	√		√			√		
Waste	√	√	√	√	√		√			
Waste water		√			√					
Emissions	√	√	√		√					
Transport					√	√				
Accidents		√	√		√	√	√			
Complaints						√	√			
Environm. training		√				√				
Environm. costs									√	√

The essential task in defining relative indicators is the selection of the reference unit or denominator. They must be precisely defined and logically related to the basic indicator.

Whenever possible, the production output derived from the input-output balance should be expressed in volume (tons). Production expressed in pieces can only be related to material input, if there is only one very homogenous product.

An alternative denominator in this case is cost of production or turnover. However, if several products have significantly different environmental impact, relating material inputs simply to total turnover (without separating turnover of the specific products) will not provide useful data.

As translation of environmental protection into cost-related figures is becoming increasingly important, EPIs have also been related to cost-related values (environmental cost indicators). This is relevant for several reasons.

In the initial phase of environmental performance monitoring data related to volumes might not be readily available; the accounting department however will be able to provide expenditures on energy consumption and disposal fees. Instead of the indicator "energy input in kilowatt hour per production output in tons" the indicator "energy expenditure in dollar per cost of production in dollar" may be used.

The cost data also helps to translate environmental performance into the "cost and savings" language that business managers understand. For example, managers who might not appreciate or react to information on the total volume of wastewater generated each year might be very interested in an estimate of the total treatment costs of wastewater each year. If an estimate of the purchase value of raw materials lost in wastewater is added, the cost information may be compelling enough to trigger action to reduce those costs, which often will also reduce environmental impact.

Likewise it is difficult to imagine the impact of 450 m^3 of hazardous waste on profit and whether it is worthwhile conducting a waste prevention study. If the same amount is expressed in waste disposal costs of €200,000 – the issue may be clearer. The data on waste disposal costs available from financial departments mostly account for the waste disposal fees only. By adding production costs of waste (storage, transportation, personnel and purchasing expenditures for the materials to be disposed) to the waste disposal fees, the necessity for cost-effective environmental protection measures becomes obvious.

Another issue needs consideration in organizations with strongly varying production and related environmental impacts. The products sold from stock in a given year may differ significantly from the quantity produced in that period. In a multi-stage production process restocking or destocking of inventory may result in significant changes of the production output. If the products differ a lot, also the environmental impact might differ significantly. As a consequence, relating materials input to turnover is of less significance and should rather be related to products produced.

For environmental performance indicators such as

- Energy input in kilowatt hour per kilogram of production
- Water input in liters per kilogram of production
- Waste categories produced in kilogram per kilogram of production

the resource input and emissions of one period should relate to the goods produced in that period. In practice, neither the quantity of products sold nor the addition of finished goods to stock are suitable reference units, as they include internal changes of stock of previous periods and purchased semi-finished and finished products. For this reason, it is recommended to use the total output of manufacturing stages as a reference unit.

For indicators related to the number of employees, the typical denominator in the service sector (e.g. waste per person), care needs to be taken to ensure clear definitions as to how the denominator is determined (part-time staff, apprentices, holidays, shift work, etc.). This is important for internal comparisons over time and for comparison of indicators between sites.

4.5.1 Specific Consumption/Eco-intensity

Eco-intensity is defined as material input in kg (absolute indicator) in relation to output in product and/or service units in kg (hectoliter, respectively), e.g. water input per hectoliter of beer production. In case of a wide range of different products, indicators for specific products and/or product groups may be calculated.

$$\text{Spec. energy input} = \frac{\text{Energy input in kWh}}{\text{Production output in kg}} = \frac{1{,}423{,}271\ \text{kWh}}{371{,}988\ \text{kg}} = 3.83\ \text{kWh/kg}$$

Generic eco-intensity indicators for most sectors are

- Raw material input in tons/product quantity in tons
- Energy input in kilowatt hour/product quantity in tons
- Water input in cubic meters/product quantity in tons
- Waste production in tons/product quantity in tons
- CO_2 emissions in tons/product quantity in tons
- SO_2 emissio ns in tons/product quantity in tons
- NO_x emissions in tons/product quantity in tons
- VOC emissions in tons/product quantity in tons
- Waste water quantity in cubic meters/product quantity in tons

Other specific consumption indicators could be input of copying paper per staff member, use of cleaning agents per square meter, or reject rate of a machine per hour of operation.

4.5.2 Eco-efficiency Ratios

By combining physical accounting data with cost data so called eco-efficiency indicators (UNCTAD, 2004) can be calculated. The World Business Council for Sustainable Development (WBCSD, 2000) defines an eco-efficiency indicator as an indicator that relates "product or service value" in terms of turnover, or profit to "environmental influence" in terms of energy, materials and water consumption, as well as waste and emission in terms of volumes.

Interpretation of these indicators requires disclosure of time series for both the nominator and the denominator. The eco-efficiency indicator can then show possible relative reduction of material input in relation to increased turnover or profit. However, as profit is influenced by other factors, like changes in world market prices and exchange rates, the interpretation of these indicators is often difficult and may have nothing to do with environmental performance. Relating material input to turnover may make more sense, as there is a direct relation to production. Examples are profit before taxes as opposed to turnover per unit water input for a brewery. Turnover would be more meaningful that profit, as is closer related to production input.

4.5.2.1 Turnover

Turnover can be very good indicator as it directly relates to production volume, which is used as the preferred reference for the material flow balance. As a physical measure from the material flow balance, the quantity of products produced and sold is the most useful denominator, preferably measured in kg, but sometimes in volume or number. If physical data are not available, turnover in monetary terms is the second best choice.

4.5.2.2 Net Sales

Net sales adjusts turnover by sales discounts, sales returns and allowances. Caution must be paid as production volumes are not directly linked to monetary sales figures, which are influenced by sales from stock, commodity prices, currency exchange rates and customer demand.

4.5.2.3 Value Added

Value added is calculated as net sales minus costs of goods and services purchased. In theory, this indicator well reflects the contribution of a company to its "products value". It is calculated by reviewing the profit and loss accounts and deducting all items comprising "purchased goods and services" from revenues. However, as value added is not mandatory for disclosure in many countries and not clearly defined from the list of accounts, its calculation requires a lot of accounting discipline and may not be generally by applied.

4.5.2.4 EBIT

Earnings before interest and taxes (EBIT) is a well know financial indicator, used as a benchmark worldwide. It is also mandatory for disclosure by for many organizations. EBIT is calculated as net sales minus all expenses, except interest and income tax. The main difference to value added is that personnel and depreciation have also been deducted from turnover.

4.5.2.5 Net Profit After Tax

This is not a good denominator for eco-efficiency rations, as the influence of financing issues significantly distorts comparison. Also differing tax laws and tax reduction provisions make interpretation difficult.

4.5.3 Percentage Distribution

A common way of presenting indicators is in relation to a baseline such as share of hazardous, municipal, and recycling waste as percent of total waste volume.

$$\text{Recycling rate} = \frac{\text{Quantity of recycled waste in t}}{\text{Quantity of total waste in t}} = \frac{3{,}461\ \text{t}}{4{,}709\ \text{t}} = 73.5\%$$

Another example would be the share of different energy carriers in the total energy input in percent, or the share of packing material weight to the total shipped product weight in percent.

Generic indicators for percentage distribution are

- Share of different materials in a product in percent
- Share of materials for product and packing in percent
- Share of products complying with defined environmental criteria (e.g. eco-labeled, organic) in percent of total products
- Share of renewable energy sources in percent of total energy input
- Share of ton kilometers on railway/ship/truck in percent
- Share of passenger kilometers of business trips by on means of transport in percent
- Share of hazardous waste in relation to total waste production in percent
- Recycling rate (share of recycled waste in relation to total waste production in percent)
- Percentage distribution of environmental costs

The annual total environmental costs assessment as described in detail in Chapter 6 and working with the excel template explained in Chapter 8 summarizes the annual

Table 4.4 Percentage distribution of environmental costs

Environmental media Percentage distribution Environmental cost categories	Air + climate	Waste water	Waste	General environmental management	Total
Material costs of non product output	14	34	20	0	68
Production costs of NPO	2	6	10	0	12
Waste and emission control costs	1	11	5	0	17
Prevention and environmental management	1	1	1	2	5
Total environmental costs	18	52	36	2	108
Environmental earnings	0	–2	–6	0	–8
Σ Total environmental costs and earnings	18	50	30	2	100

costs to a one page template which automatically transforms into a percentage distribution of the annual environmental costs of the previous business year. The columns show the distribution of costs by environmental media affected while the lines display the distribution within the environmental cost categories. The example in Table 4.4 from a case study in the pulp and paper industry in Austria (Jasch and Schnitzer 2002), shows that water management accounts for up to 50% of all environmental cost, with waste and air/climate accounting for about 30% and 20%. But wasted material input is by far the dominant cost factor, accounting for about 80% when calculated by cost categories (material purchase and processing costs). For other sectors and regions, these shares will vary.

4.6 Calculating Savings Based on Performance Indicators

A brewery in Austria, Murauer Bier, installed an Environmental Management System (EMS) in 1995, based on the European Union's 2001 Regulation on Environmental Management and Audit System. Murauer's EMS is supplemented by an extensive system of environmental performance indicators (Murauer Website). The company uses physical and monetary accounting data to calculate these EPIs and to calculate the annual monetary savings achieved since the implementation of the EMS.

Absolute EPIs calculated by Murauer include the total amounts of all significant material inputs (for example, hectoliters of fresh water, and kilograms of heating oil). Relative EPIs are also created by calculating the ratio of each Materials Input to hectoliters of Product Output, that is, beer. Similar absolute and relative EPIs are calculated for the brewery's Non-Product Outputs (for example, glass, paper, wastewater, carbon dioxide and other air emissions). The brewery Murau also compares

EPIs from year to year to track its environmental performance trends and overall progress. The following EPIs illustrate the success of some of Murauer's waste minimization efforts during a 5-year time period:

- Reduction in fresh water use per unit product – 9%
- Reduction in fuel oil use per unit product – 30%
- Reduction in wastewater generation rate per unit product – 32%

Monetary savings are calculated separately for each Material monitored on the Input Side of the material flow balance. The first year of the establishment of the indicator system is taken as reference. Each year, the reduced material input due to efficiency gains is multiplied with current prices, thus indicating annual savings. These efforts saved the medium-sized firm approximately US$186,000 in 5 years (Jasch and Schnitzer, 2002, Murauer Website, Environmental Statement of Obermurtaler Brauerei 2003).

Chapter 5
Environmentally Relevant Equipment

Chapter 5 describes the different types of environmentally relevant equipment, which is often the first step when conducting an EMA assessment. The term "equipment" may comprise a single machine or an entire production hall, but the assessment is best performed on a cost center level. In order to provide the necessary data for investment appraisal, actually three categories of environmentally relevant equipment should be distinguished:

- *End-of-pipe equipment for treatment of waste and emissions*
- *Integrated cleaner technologies which prevent emissions at source*
- *Scrap producing equipment and energy conversion losses*

The different approaches of IFAC, UN DSD and UNIDO in opposition to SEEA and CEPA regarding the inclusion of cleaner technologies and integrated prevention are highlighted.

The first step when conducting an EMA assessment often is defining environmentally relevant equipment. From the point of view of cost assessment it is advisable to check the list of cost centers and investigate to what degree these can be evaluated for their environmental relevance. In many organizations it will be advisable to define the environmentally relevant equipment based on the cost centers. While some, like the waste water treatment plant or waste disposal dumps, will be 100% environmentally relevant and posted to cost centers on their own, for most other production equipment the environmental share needs to be estimated, if significant, as it is probably not possible to separate the environmental part of the production technology from the overall purchase price and related depreciation.

Investments are capitalized and accounted for by depreciation in the profit and loss account if they bear a future benefit, otherwise they are immediately expensed. As a rule, expenses which do not lead to future economic cost savings should be expensed in the year in which they occur. End-of-pipe technologies qualify as assets as they are often an easy, though expensive solution to fulfill

C. Jasch, *Environmental and Material Flow Cost Accounting*: Principles and Procedures, Eco-Efficiency in Industry and Science 25,

legal compliance. Their value can easily be determined as they are typically stand-alone treatment facilities.

For any type of activity and especially for investments which, in addition to their primary purpose (usually an economic one), also have an environmental aspect (and vice versa), the question arises if the equipment is environmentally relevant as well and to what degree. While most equipment will be environmentally relevant, as it consumes resources and energy and produces waste in addition to products, only environmental protection equipment should be externally reported to statistical agencies etc. Equipment producing significant amounts of waste and emissions however also needs monitoring as it provides improvement potentials with environmental as well as economic benefits.

To help in determining if the equipment was purchased for production or for environmental protection, imagine the equipment in an area where there are no environmental laws or no people living and investigate, whether it is needed for production or not. The question, whether equipment was installed in order to fulfill legal requirements or not doesn't allow a precise distinction between environmentally relevant or production oriented, as legal permits are different in each country. In addition, legal requirements normally don't request a certain technology but a certain emission level, which can be obtained by several technological options.

It is not advisable to spent too much time on trying to trace the environmental investments of previous years if they have been included into general cost centers (like "building") and the related depreciation can not be assessed easily. It is more advisable to develop a procedure to make sure that in the future environmentally relevant equipment is flagged already at the point of time of posting it to a cost center and starting depreciation. Often the environmental manager should be consulted for the decision, whether an equipment is environmentally relevant and to what degree. Flagging in the organizations information system allows later assessment of investment volume and annual depreciation.

Equipment Depreciation distributes the investment costs of equipment over its expected lifetime, recorded on an annual basis. The amount of depreciation should follow the principles of accounting that the organization normally applies. The following models are possible:

- Depreciation from financial accounting is used; once the equipment is depreciated, there is no more annual expenditure for depreciation.
- Depreciation can also be calculated based on cost accounting principles. The value of depreciation can be taken from financial accounting, but may be continued, once the equipment is depreciated in financial accounting.
- Cost accounting also allows to calculate the depreciation on the basis of the expected new purchase price for similar equipment and to add calculatory interest.
- In the case studies a pragmatic solution often applied was to calculate depreciation on the basis of the actual investment costs but to simply distribute the depreciation over a time span of 10 years for all environmentally relevant equipment and not to relate back to the asset accounting system in later years. The reason for this approach is that often the investment volumes can be traced back from

project accounts, while the actual posting in the asset accounting system remains a mystery.

In accordance with financial accounting rules an investment should be recorded at the time of put in function and not during the project development phase (cash outflow of the company). In many organizations the recording for environmental statistics in done by the environmental manager, who has no access to the corporate accounting system. He therefore tends to report investments at the stage of projects, which show the annual cash outgo, spend for these investments, but differ from the treatment in the accounting system, which records a project only at the put in function stage, which is also the point of time, when deprecation starts. Some countries, e.g. Rumania, explicitly ask for the cash outgo in a given year which conflicts with the set up of the corporate wide accounting system, which flags environmental investments at the time of put in function, records the investment volume at this point of time, and lists related investment grants. The depreciation in future years is thus automatically calculated.

Investments that have been considered as environmentally relevant will thus automatically be included with their related operating costs in the upcoming years. In many organizations, this data is taken directly from cost center reports, which collect depreciation, operating materials, services and personnel for a defined cost center. Equipment that has been defined as environmentally relevant consequently should be reported with its operating costs also in the following years.

Different levels of national environmental production standards may also cause a question. International corporations have been faced with the fact, that the same technology can be treated as integrated prevention in one country and state of the art in another. This has led to the situation, that the same technology is treated different in each country. In company projects this was accepted, as in many countries technologies, which are state of the art in the European Union, are requested by environmental ministries elsewhere and clearly qualify as mandatory environmental protection there.

5.1 Classification of Environmentally Relevant Equipment

The IFAC EMA guidance document clearly separates between equipment for treatment and prevention. As the focus is on annual costs, annual depreciation is collected for total annual costs but in the assessment template developed for the data assessment (see Chapter 8.1) the annual investment volume is collected as well.

But in order to provide the necessary data for investment appraisal, actually three categories of environmentally relevant equipment may be distinguished (Jasch and Schnitzer, 2002):

- End-of-pipe equipment
- Integrated cleaner technologies
- Scrap producing equipment

While only end-of-pipe and integrated prevention technologies, depending on the reporting purpose, qualify as environmental protection equipment, also equipment producing significant amounts of waste and emissions needs to be monitored within environmental management and in order to provide the data necessary for investment appraisal (Staniskis et al. 2005).

5.1.1 End-of-Pipe Equipment

Investments which are incurred solely for the purpose of emission treatment are typically end-of-pipe technologies, i.e. devices which are installed for cleaning purposes after the production processes. Filters, waste collection equipment and waste water treatment plants are common end-of-pipe-technologies which help to concentrate or hold back toxic substances. However, they usually do not solve the problem at source, but rather prevent uncontrolled release in exchange for controlled release.

End-of-pipe equipment comprises equipment, machines, constructions, etc. that exist solely for environmental protection or clean up, and are not necessary for production (e.g. wastewater treatment, dust removal filters, waste separation and compression equipment, sound insulation walls, etc.). This equipment is 100% environmentally relevant. It requires investments, causes operating costs (personnel and operating materials), and needs to be maintained. This equipment is often monitored on separate cost centers, from which the personnel-, and other operating costs can be traced.

A clear distinction between environmental and production equipment is often only possible for investments in end-of-pipe technologies which, however, are unable to fully address an emission problem but usually only transpose it to another environmental medium (e.g. from air to soil) and help to fight the symptoms, but not the cause of pollution.

Examples of **waste and emission control equipment** include

- Waste handling equipment (such as solid waste dumpsters, waste transportation equipment)
- Waste and emissions treatment equipment (such as wastewater treatment systems, air Scrubbers
- Waste disposal equipment (such earth moving equipment for an on-site landfill)

Waste and Emission Control systems include both standalone, "end-of-pipe" control equipment, where the sole purpose is to control waste and emissions, as well as integrated control equipment, which may be closely integrated into actual production equipment. Organizations with large, standalone waste and emission control equipment, such as wastewater treatment plants, often record cost information related to the operation of this equipment in separate cost centers within their accounting systems.

In such cases, many of the associated Waste and Emission Control Costs can be taken directly from these cost center reports.

5.1.2 Integrated Cleaner Technologies

Energy and Environment have been crucial issues of UNIDO's work for over 30 years. Within the Organization's endeavors to make modern energy accessible to developing countries, to establish renewable energy resources and to foster environmental sustainability, UNIDO supports cleaner and sustainable production. The UNIDO Cleaner Production programme (CP) aims at building national CP capacities, at fostering dialogue between industry and government and at enhancing investments for transfer and development of environmentally sound technologies. Thereby the Organization attempts to bridge the gap between competitive industrial production and environmental concerns.

UNIDOs webpage (www.unido.org/cp) defines cleaner production as a preventive, integrated strategy that is applied to the entire production cycle to

- Increase productivity by ensuring a more efficient use of raw materials, energy and water.
- Promote better environmental performance through reduction at source of waste and emissions.
- Reduce the environmental impact of products throughout their life cycle by the design of environmentally friendly but cost-effective products.

The definition of Cleaner Production that has been adopted by UNEP and is most commonly applied worldwide reads as (www.uneptie.org/pc/cp/understanding_cp/home.htm).

Cleaner Production (CP) is the continuous application of an integrated preventive environmental strategy to processes, products, and services to increase overall efficiency, and reduce risks to humans and the environment. Cleaner Production can be applied to the processes used in any industry, to products themselves and to various services provided in society.

- For **production processes**, Cleaner Production results from one or a combination of conserving raw materials, water and energy; eliminating toxic and dangerous raw materials; and reducing the quantity and toxicity of all emissions and wastes at source during the production process.
- For **products**, Cleaner Production aims to reduce the environmental, health and safety impacts of products over their entire life cycles, from raw materials extraction, through manufacturing and use, to the 'ultimate' disposal of the product.
- For **services**, Cleaner Production implies incorporating environmental concerns into designing and delivering services.

Environmental sound technologies (EST) protect the environment, are less polluting, use all resources in a more sustainable manner, recycle more of their wastes and products, and handle residual wastes in a more acceptable manner than the technologies for which they were substitutes. EST in the context of pollution are "processes and product technologies" that generate low or no waste, for the prevention of pollution. They also cover "end-of-the-pipe" technologies for treatment of pollution after it has been generated.

Integrated technologies comprise equipment, machines and constructions for production purposes that produce less waste or emissions than previous technologies (enameling line with after-burning, boiler plant with flue gas cleaning, bottle washing line with separate discharge of glass, paper, and metal, all equipment capsuled for noise reduction or with integrated water reuse systems, etc.). In many cases it is possible to minimize waste and emissions b using equipment with integrated pollution prevention and control. Sometimes this equipment is more expensive at purchase, but often also more economical during production. An example of such equipment would be a comparatively more expensive enameling line that sprays more efficiently, which means higher depreciation costs, but also lower material inputs and waste.

Integrated prevention can also be related to reducing the environmental impacts of products. An example of such technologies would be the desulfurization of gasoline to reduce environmental impact at combustion.

The proportion of environmentally relevant investment depends on the increase in the investment costs in comparison to state of the art technology and whether the motivation for the installation of the equipment was influenced significantly by environmental considerations. There is no clear definition for the environmental share of integrated technologies and often they reflect state of the art technology and should not be treated as "environmental investment". If the additional costs were significant, their magnitude and/or the percentage of the investment costs may be estimated and the related depreciation recorded. If the operating costs are significant and can be taken from the cost center reports, it should be done. In other cases it may be sufficient to record the material inputs under the cost category for non-material-output and omit the related personal costs.

Some **equipment used for Prevention and Other Environmental Management** can be stand-alone equipment (such as a new computer system for environmental data collection). The annual depreciation costs for such equipment would be included under this cost category. Other equipment used for Prevention may be closely integrated into production equipment (such as a solvent distillation and re-use system that is an integral and automated part of a chemical manufacturing process). In other cases, equipment (for example, a high efficiency paint spray gun) may simply contribute to Preventive Environmental Management because it inherently uses energy or raw materials more efficiently and produces less waste than alternative equipment. In such cases, an organization may wish to estimate what percentage (if any) of the annual depreciation costs for the equipment should be designated as "environment-related." This estimate might be based on a consideration of the primary reasons for purchasing that particular piece of equipment, for example, for environmental or materials efficiency considerations.

5.1.3 Scrap Producing Equipment and Energy Conversion Losses

Since producing waste and emissions is environmentally relevant, so is equipment, which produces them. Scrap producing equipment comprises the share of equipment that produces scrap, waste and emissions (e.g. old boilers with inefficient power conversion, non-insulated pipes that cause avoidable energy losses requiring higher energy input, enameling lines that produce painted products that have to be painted again, steam supply with heat losses, all equipment which produces products with insufficient quality, etc.).

This equipment should not be listed as investment in environmental protection, but its recording together with the related material and personal input is necessary to provide sound data for investment appraisal. If the actual consumption, losses and related costs of existing technologies are not known, applying investment appraisal to calculate the savings from installing new technologies can not provide reasonable results.

The environmentally relevant portion of this equipment may be defined by the share of losses, scrap and waste in relation to high quality product output. If this share is significant, the related depreciation may be recorded in a separate category to have the data ready for decision making. The material losses will be recorded under the NPO section. Sometimes it may also be advisable to trace the operating costs from the cost center reports of the scrap producing equipment.

The environmentally relevant portion of equipment that converts energy (boiler plants, transformations, pressure reduction plants for natural gas, air compressors, air conditioning, etc.) depends on the portion of lost energy. There are four approaches to evaluating the energy use:

1. Evaluating energy as non-product output (NPO): Since energy does in most cases not enter the product, but is a typical operating material escaping as heated water, air, and radiation, it is considered as 100% NPO. Recording total energy input as operating material and neglecting depreciation of equipment with energy losses allows for the best possible consistency with the input-output mass balance, and is the approach taken in most companies. The data collection can thus continue without further technical estimation.
2. Evaluating energy conversion losses: Since energy is required for most production processes, it may be reasonable to only regard the transformation and transportation losses (combustion losses, pipe losses, etc.). The related efficiencies may be known (e.g. with combustion) or have to be estimated (e.g. propulsion, conduction, etc.).
3. Evaluating avoidable losses: Since energy losses are not completely avoidable, the evaluation can be calculated based on the difference between the current

system and the state of the art. If there are more efficient systems available, than the difference is environmentally relevant. For example, one can compare the current fleet of cars to the most fuel efficient vehicles available. Option 2 and 3 should however only be undertaken, if the information produced is relevant for decision making (e.g. for benchmarking technologies or sites).
4. Evaluating the energy use of the environmentally relevant equipment: The energy use of environmentally relevant equipment (e.g. compressors, waste water plants, after burners, etc.) is, just as the other operating costs of such equipment, always 100% environmentally relevant.

The approaches may be combined. In several company projects, the energy input recorded on the environmentally relevant cost centers is recorded under control costs (and to a very small degree under prevention costs). For on site energy production and product production equipment energy is either recorded with 100% purchase costs or based on conversion losses, sometimes only in relation to state of the art equipment and posted under NPO. Energy used for transport, heating and lightening is calculated with 100% of purchase costs as NPO.

5.2 Environmental Investments According to SEEA and CEPA

Two types of capital expenditure are distinguished in SEEA (2003, p. 215):

Expenditure on **end-of-pipe technologies** "used to treat, handle or dispose of emissions and wastes from production. This type of spending is normally easily identified even within the context of ancillary activity because it is usually directed toward an "add on" facility which removes, transforms or reduces emissions and discharges at the end of the production process."

Expenditure on **integrated investments**, also called cleaner technologies. "These are new or modified production facilities designed so that environmental protection is an integral part of the production process, reducing or eliminating emissions and discharges and thus the need for end-of-pipe equipment."

SEEA states (p. 215): "Integrated investments may result from the modification of existing equipment for the explicit purpose of reducing the output of pollutants, or from the purchase of new equipment whose purpose is both industrial and for pollution control. In the first case, expenditure can be estimated from the cost of the modification of existing equipment. In the second, the extra cost due to pollution control has to be estimated; that is, the cost of non-polluting or less-polluting. Equipment is compared to that of "polluting or more polluting" reference equipment."

Such estimates are difficult to make when reference equipment no longer exists or new equipment presents other advantages in addition to its beneficial effects on the environment. These may include savings or substitution of raw materials, higher productivity and so on which cannot be isolated in terms of cost. The difficulty arises because the steady integration of environmental standards in equipment and processes means that eventually it becomes impossible to identify a part of the expenditure as environmental. Given the different speed at which new environmental standards are incorporated into different types of equipment and in different countries, comparison of long time series across industries and countries is difficult. However, a misleading picture is obtained if the cost of significant capital equipment is ignored.

SEEA requests to make a clear distinction between purpose and effect. For example, in the case of environmental protection, actions undertaken for other than environmental purposes can have positive environmental effects (for example new technologies may lead to reductions in energy use, material consumption and discharges to the environment), whereas it is conceivable that actions undertaken with an environmental protection purpose may not actually have a beneficial environmental effect. But only the "environmental purpose criterion" is applied to qualify an environmental investment!

The CEPA definition which request that measures undertaken for cost saving reasons are excluded from environmental expenditure is not only difficult to understand from a corporate perspective, but also poorly defined. In corporate accounting companies often specify a required return of investment period (e.g. 3–4 years) and allow for longer periods for environmental protection equipment. CEPA doesn't specify, if a technology that falls out of the standard corporate investment pay off period, but eventually will pay off, could qualify as environmental investment (likewise Sprenger, 2007). This is not suggesting that SEEA should define pay off cycles but rather demonstrating that the criterion is not practical.

When comparing the definitions for pollution prevention and cleaner production of IFAC und UNIDO with the SEEA approach, it is important to notice that SEEA

- Does not include measures to reduce the input of materials, energy and water and increase resource efficiency
- Does not include measures for energy efficiency and renewable resources as they would qualify under "resource management"
- Does not allow for measures which have a positive pay back
- Does not allow for measures, where the primary purpose is not environmental protection but resource and production efficiency
- Does not allow for measures related to reduction of the environmental impact of products

The questionnaires send out by national statistical agencies often don't take over these restrictions but allow for more flexible interpretation of cleaner technologies. Under the current revision of SEEA these issues are addressed.

Chapter 6
Monetary Information

Chapter 6 describes the different environmental cost categories in detail. They are based on the classification in the IFAC EMA Guidance Document. For each cost category the sub-categories relating to financial accounts, such as equipment depreciation, operating materials, water, energy and personnel are discussed and examples provided. In addition, environment related earnings from grants for investments or from scrap sales are described. National statistical institutes require reporting of environmental costs by the environmental domain affected. The chapter concludes with a case study of the pulp and paper company SCA Laakirchen, which shows the average percentage distribution of the previously described environmental cost categories.

Similar to the physical information collected under EMA, monetary data can be collected for an organizaztion as a whole, or for particular sites, input materials, waste streams, process or equipment lines, product or service lines, depending on the intended use of the information (for example, process optimization, investment appraisal, or assessment of total annual costs). As a starting point, setting up a system to record annual environmental costs is recommended, which can later be refined for more detailed assessments. Many organizations have also established internal management standards with procedures and responsibilities for the consistent recording of environmental data and costs.

Even though the material flow balance and monetary information are being presented separately, it is essential to link all physical Inputs and Outputs with the appropriate cost categories for consistent and accurate EMA. All cost categories besides Research and Development link back to the physical accounting information discussed in Chapter 3: Raw and Auxiliary Materials, Packaging Materials, Operating Materials, Water, Energy, Product Outputs and Non-Product Outputs.

Some costs may fit into more than one of the cost categories listed below. For example, the purchase costs of operating materials used to run waste treatment equipment should be recorded under Waste and Emission Control Costs, if the waste water treatment plant has been installed as a separate cost center and the use

C. Jasch, *Environmental and Material Flow Cost Accounting*: Principles and Procedures, Eco-Efficiency in Industry and Science 25,

of chemicals and other operating materials has been posted there. All costs that directly relate to equipment defined as environmentally relevant with a certain percentage and available as separate cost center should be quoted in the cost categories Waste and Emission Control or Integrated Prevention.

However, most materials will probably not be available from cost center reports. Therefore, in the cost category of NPO the materials consumed and available from the profit and loss accounts are listed and quoted with their measured or estimated loss percentage. When assessing total annual environment-related costs, materials quoted from the cost center reports and posted under Waste and Emission Control Costs thus need to be deducted from the same cost subcategory under NPO.

6.1 Overview on the EMA Cost Categories in the Excel Template for Total Annual Environmental Costs

For the UN DSD Working Group and the IFAC EMA standard a set of cost categories was developed which

- Reflects the different nature of environmental burden (material and product related costs, end-of-pipe control costs as well as integrated prevention and environmental management).
- Specifies the subcategories by traditional accounting terminology (such as depreciation, materials, labor).
- In accordance with the requirements of statistical agencies distributes the costs by environmental domain affected (such as air, water and waste).

IFACs cost categories are shown in Table 1.2 and 2. More specific descriptions of the categories and types of costs are given later in this chapter. To assist the cost assessment an excel template has been developed that is available for download at www.ioew.at and. www.springer.com/978-1-4020-9027-1

IFACs cost category 1–Material Costs of Product Output–are assessed directly together with the material flow balance, as shown in Table 2.1. All the materials input is recorded simultaneously in physical as well as monetary terms in order to ensure consistency.

IFACs cost category 6–Less Tangible Costs–are not accounted for, as they cannot be traced from the accounts, but need to be estimated separately. They are not part of annual expenditure of the previous business year, but a point of consideration for future oriented decisions including investment appraisal. Examples of Less Tangible Costs related to the environment include: liability (such as legal judgments related to natural resource damage); future regulation (such as likely future costs of stricter regulation of greenhouse gas emissions); productivity (such as worker absenteeism due to pollution-related illness); image and stakeholder relations (such as, barriers to financing for projects with negative environmental components); and externalities (external effects on society, such as the loss of property values due to proximity to highly polluting factories).

Table 6.1 Environment related costs and earnings assessment template

Environment-Related domains/cost categories	Air and climate	Waste water	Waste	Soil, groundwater and surface water	Noise and vibration	Biodiversity and landscape	Radiation	Other	Total
Materials costs of non-product outputs									
Raw and auxiliary materials									
Packaging materials									
Operating materials									
Water									
Energy									
Processing costs									
Waste and emission control costs									
Equipment depreciation									
Operating materials									
Water and energy									
Internal personnel									
External services									
Fees, taxes and permits									
Fines									
Insurance									
Remediation and compensation									
Prevention and other environmental management costs									
Equipment Depreciation									
Operating Materials, Water, Energy									
Internal Personnel									
External Services									
Other Costs									
Research and development costs									
Environmental costs									
Earnings									
Total environmental costs and earnings									

Table 6.1 shows the environmental cost assessment template developed for EMA for the assessment of total annual costs. It can be adopted to fit company needs. This chapter provides information on the different categories. The annex provides checklists for determination of environmental costs by environmental media. Chapters 7, 8, and 9 explain how to perform an EMA assessment by using the EMA Excel Assessment Template in more detail.

6.2 Distribution by Environmental Domain

This section explains the distribution of environmental costs by environmental domain affected in accordance with SEEA requirements. The columns in Table 6.1 show the assignment of environment-related costs to environmental domains. These are a modified version of the domains that European statistical offices must use in reporting businesses' environmental protection expenditures to Eurostat, the statistical arm of the European Commission. The national statistical offices collect the required information directly from businesses. The member countries of the Organization for Economic Co-operation and Development (OECD) also use the European Commission domains, as does the System of Integrated Environmental and Economic Accounting (SEEA) of the United Nations:

The classification that SEEA (2003) suggests for organizing environmental protection activities is the Classification of Environmental Protection Activities (CEPA). This classification applies to environmental expenditures and products. The environmental domains of CEPA are classified as

- Protection of Ambient Air and Climate
- Wastewater Management
- Waste Management
- Protection and Remediation of Soil, Groundwater and Surface Water
- Noise and Vibration Abatement
- Protection of Biodiversity and Landscape
- Protection against Radiation
- Research and Development
- Other

To provide for maximum consistency with existing international approaches, this classification has also been used for the EMA environmental cost assessment scheme, with the exemption of research and development activities, as they are covered as separate cost category. If appropriate, organizations might also want to consider adding a column for health and safety issues, for product oriented prevention activities or for other related issues. Some organizations, e.g. utilities like the Austrian Verbundgesellschaft, have, together with the Austrian environmental protection agency, developed a more detailed categorization specifically for each business group. Several organizations have also omitted the categories for noise, biodiversity and radiation, as they may not be relevant for the business area.

Within the CEPA, environmental protection activities are first classified by environmental domain (air, waste, nature protection, etc.) and then by type of measure (prevention, treatment, etc) (SEEA, 2003, p. 201). It is important to notice, that for the EMA assessment the environmental costs are **first assessed by standard accounting categories** and **only then assigned to environmental domains affected**. This way it is the task of the accountants with support from the environmental manager to set up a consistent and complete data information system.

The CEPA approach and the questionnaires send out by statistical agencies tend to go the other way round. They ask for environmental domains and thus the questionnaire is being answered by the environmental manager who often has no direct access to the accounting system and no overview on the total corporate cost structure. The information reported is thus often not complete and consistent.

Environmental domain categories are useful not only for compliance with external reporting requirements, but also can show interesting and useful results and trends for internal management purposes. The most widely used application is benchmarking environmental costs by domain from year to year and among multiple sites of corporations, as is illustrated by the case study of the pulp and paper plant SCA Laakirchen in Chapter 6.8 and by the case study from DANISCO in Chapter 7.9.

6.3 Material Costs of Non-product Output

The material flow balance as described in Chapters 2 and 3 assesses all materials inputs and resulting product and non-product outputs of organizations in the manufacturing sector. On the material input side, physical and monetary values are collected simultaneously to ensure consistency of data. On the output side of the material flow balance, Product Outputs usually make up the biggest amount of physical outputs from manufacturing operations. But, the total NPO (Waste and Emissions generated in manufacturing) can still be quite large, costly and environmentally significant. In operations where there is no physical product, all Input Materials leave the organization as NPOs, by definition.

This cost category covers the purchase costs of all materials not converted into a Product but into Non-Product Output (Waste and Emissions). For Materials where actual consumption is monitored by stock management, not the value for materials **purchased**, but **consumed for production** is used respectively.

The costs of treating or disposing of those Waste and Emissions are considered separately in a different cost category. The physical accounting side of EMA provides the information on the amounts and flows of materials needed to assess these costs.

Once the input side of the material flow balance has been assessed, for the cost category Materials Costs of NPO for each inputted materials loss percentages need to be measured or estimated. These losses are recorded with the related input prices for Raw and Auxiliary Materials; Packaging Materials; Operating Materials; and sometimes Water and Energy.

Although many organizations may consider these costs to be related to efficiency or quality, they are also environment related as the physical part of lost materials constitutes Waste and Emissions and the financial part helps an organization to cost-effectively manage the environmental impacts of its Waste and Emissions. By visualizing these costs, organizations might become interested in acquiring more efficient process equipment that generates less waste per unit product output.

Whatever has not left the company as a product is a sign of inefficient production and must by definition be waste and emissions. Determining the material flows for, at least, raw and auxiliary materials is therefore imperative for environmental cost assessment. The material purchase cost of wasted materials is the most important environmental cost factor for manufacturing companies, accounting for 40–70% of total environmental costs, depending on the value of raw materials and the labor intensity of the sector.

Not all types of waste and emissions can be reduced but it is clearly in the financial best interest of organizations to use as little materials, energy and water as possible. Current ZERO waste initiatives at least try to make sure that all NPO is transformed into a by-product, that can be used for other processes, or into an output that has no negative impact on the environment. Preventive and proactive environmental management that reduces the amount of waste generated, rather than just treating the waste once it is generated, can reduce not only the purchase costs of materials lost as wastes, but also subsequent waste control and treatment costs. Thus, assessment of these costs also allows managers to better assess the potential monetary value of preventive environmental management.

6.3.1 Estimating Loss Percentages

In even the most efficient manufacturing operations, some Raw and Auxiliary Materials and Packaging Materials will not be converted into Product Output, but become NPO. Operating Materials, Water, and Energy never intended to become part of a physical product will also become NPO by definition.

The purchase costs of merchandise can also be tracked if significant amounts of merchandise become waste before they are sold, for example, the waste generated at repackaging for different countries or due to spills and damage in the warehouse.

Non-product raw material output will mostly be disposed of as solid waste. Only in those rare cases where the company's product is gaseous (industrial gases, perfume), will it be found in the air. More common is a liquid product (beer, milk) that goes down with wastewater in which case only a certain percentage of water input should be quoted under purchase value of non product output.

For most organizations energy is not a product but an operating material and thus should be recorded with its total purchase value. However, in some organizations it makes sense to quote the conversion losses or to benchmark against best available technology. These options are described in Chapter 5.

In several organizations raw material losses are monitored by quality management and the data is used for controlling purposes. However, not for all material groups measurements will most likely be available at the start of an EMA project. For a first estimate, calculations for scrap percentages can be used to estimate the NPO of the different material subgroups. Experience shows, that the people working in production and the environmental manager can provide much more accurate estimates for loss percentages for the different material groups than the accounting departments. Several case studies have revealed that communication between production and financial departments on actual loss percentages should be improved and that accounting departments tend to calculate with rather outdated estimates, while the production departments don't automatically share their monitored data with controlling. Eventually, with more detailed material flow balances, scrap percentages may need adjustment. The reasons, why materials do not become products are manifold and well worth study.

Product returns, obliteration, repackaging for other countries or specified customer requests, quality control, production losses, spoilage, wastage, decay in storage, shrinkage, etc. are some of the causes of waste generation that call for measures to increase production efficiency, which may be profitable both from an economic and environmental point of view.

6.3.2 Calculating Processing Costs of NPO

Business economics distinguishes three production factors: materials, capital (equipment, related annual depreciation and financing cost) and labor. Waste and Emissions (synonym with NPO) not only carry material purchase prices, but may be calculated with their respective production costs. NPO has also undergone processing in the company before leaving it again. Thus, wasted labor and capital costs may be added.

Work time lost due to inefficient production, and a share of depreciation for equipment as well as possible other costs, like financing costs, may be accounted for. For waste of raw materials and products in the various phases of production (usually solid or liquid) pro-rata production costs are mostly calculated as a percentage based premium on the material purchase value.

When estimating Materials Processing Costs of NPO, care must be taken to avoid double counting. In most organizations, the percentage based premium for the calculation of production costs probably includes not only equipment depreciation and personnel costs, but also costs already covered by other categories. In several case studies production costs of NPO have thus only been calculated on product losses by the cost accountant himself. Calculating processing costs of NPO is not a prerequisite for EMA but rather an interesting add-on for the controller or cost accountant and requires very detailed knowledge on the organizations set up of the system design.

6.4 Waste and Emission Control Costs

This category covers the costs of handling, treating and disposing of Waste and Emissions (equipment depreciation, operating costs, external services, disposal fees, etc.); remediation and compensation costs related to environmental damage; and any regulatory compliance costs related to Waste and Emission control. Insurance and provisions for environmental liabilities also reflect the spirit of treatment instead of prevention.

The costs of controlling and treating all forms of Waste and Emissions once they have been generated are being collected. The related control activities include equipment maintenance; internal waste handling; waste and emission treatment; off-site recycling; waste disposal; remediation of contaminated sites and other pollution clean-up. These costs are not related to the production process as such but deal with "unwanted" output once it has been generated. It is in the best interest of an organization to minimize these costs by preventing the generation of waste and emissions at source and at the same time maintaining a high level of environmental performance. Prevention activities are covered in the next cost category.

This cost category includes costs for waste and emission control and treatment and comprises the following sub-categories:

- Equipment depreciation for end-of-pipe technologies
- Operating materials; water and energy
- Internal personnel
- External services
- Fees, taxes and permits
- Fines
- Insurance and
- Remediation and compensation

6.4.1 Equipment Depreciation

Waste and emission control equipment includes typical end-of-pipe treatment facilities which have been added at the end of the production process to reduce the environmental impact of waste and emissions. More explanation on the different types of environmentally relevant equipment is provided in Chapter 5. Examples of waste and emission control equipment are

- Waste handling equipment (such as solid waste separation, transport and compression equipment)
- Waste disposal equipment (such as equipment installed on a company owned on-site landfill)
- Waste and emissions treatment equipment (such as wastewater treatment systems, flue gas desulfurization and NOx removal, noise abatement installations)

Most of this equipment will be stand alone, end-of-pipe control equipment. Large, standalone waste and emission control equipment, such as wastewater treatment plants, are often recorded in separate cost centers. In such cases, the associated Waste and Emission Control Costs can be taken directly from these cost center reports.

But some of this equipment may be closely integrated into actual production equipment (and the depreciation may thus not be recorded separately on the cost centers). The percentage share of investment costs that relate to relate t environmental protection thus needs to be estimated and should be recorded if significant. For all other waste and emission control equipment that does not have separate cost centers, an organization will need to spend some time tracing at least the former investment costs in order to estimate the annual depreciation.

6.4.2 Operating Materials, Water and Energy

As stated previously, Operating Materials are Materials Inputs never intended to leave the organization as a product or together with the product, but are still necessary to run the organization. They include water and energy. An example of an Operating Material used specifically for the purpose of Waste and Emission Control are the chemicals used in an on-site wastewater treatment plant.

Once waste and emission control equipment has been defined, the annual costs for related operating materials, maintenance, inspection etc. can often be recorded directly from the related cost center reports. If this is the case, than the related amount taken from the cost center report must be deducted from the operating materials recorded in the cost category of NPO. To the degree possible, the information available in the cost center reports of equipment defined as waste and emission control equipment should be recorded in this cost category and not under general NPO.

6.4.3 Internal Personnel

Personnel dealing with waste and emission control should be recorded in this cost category. Again, most of the data will be taken directly from the cost center reports of the related equipment defined in Section 6.4.1. It mainly applies to the personnel of waste collection departments, and the people in charge of wastewater and air emission control, dealing directly with the identified waste and emissions streams and equipment. Examples include internal personnel for

- Maintenance (such as wastewater treatment plant maintenance)
- Waste handling (such as waste segregation, collection, testing, internal transport)
- Waste and emissions treatment (such as operation of wastewater treatment plants and incinerators)

- Waste disposal (such as management of an on-site landfill)
- Regulatory compliance (such as monitoring, record keeping, inspections, notification and training)

Internal Personnel costs can either be calculated with the values available from cost center reports. In addition it is advisable to estimate the average annual time spent by the related people involved and to calculate the related costs based on average personal costs for this qualification level. It is not necessary to record the actual salaries of the personal involved.

6.4.4 External Services

The costs of all External Services provided by consultants, contractors, law firms, etc., related to Waste and Emission Control should be included here. Again the cost center reports for the waste and emission control equipment may contain some of this information. However, experience shows unless this data is recorded on defined cost centers it is practically impossible to later trace the related invoices. It is recommended to have the environmental and production manager consider the services used in the previous business year and than try to trace at least some of the related costs from the related accounts. Sometimes the supplier accounts offer a better source of information than the profit and loss accounts.

6.4.5 Fees, Taxes and Permits

This category includes any Fees, Taxes and Permits related to Waste and Emission Control. SEEA (2003, p. 242) distinguishes environmental taxes from fees for a service. Fees are included in environmental expenditure while taxes are not. The charges paid by households and businesses for a variety of services including the provision of piped water and the collection of refuse have in recent years been separated from general government services and the charges made to households and businesses are being regarded as payments for a service rather than a tax. Fees which represent a payment for a service should be covered in the environmental treatment expenditure.

Examples for treatment fees and permits include

- Packaging license fee
- Waste disposal fee
- Waste water treatment charge
- Water withdrawal charge
- Parking allowance
- Highway allowance

- Environmental production permits
- Greenhouse gas emission permits

OECD, Eurostat, the IEA and the European Commission's Directorates General for Environment and Taxation have developed a statistical framework on environmental taxes (Eurostat, 2002). The framework provides the following definition of environmental taxes: "a tax whose tax base is a physical unit (or a proxy of it) that has a proven specific negative impact on the environment" (SEEA, 2003, p. 246).

The role of taxes in the management of environmental resources is to increase the price of the products or the costs of production of the activities concerned. Sometimes the revenues from such taxes may be designated to remedy particular forms of environmental damage. These taxes are not considered part of environmental expenditure by SEEA (2003). From a corporate point of view fee and taxes are being collected on the same account and it is not practical to separate them later. But as not many countries are applying environmental taxes the way defined by SEEA it doesn't really matter so much. In addition the basis for the identification of environmental taxes will often be tax revenue statistics and not corporate assessments.

Examples include

- Energy products
- Vehicle tax
- Taxes on emissions to air or water

Many of these taxes are charged when selling the product and thus included in the product price. Companies (the users) don't record them on separate accounts and their assessment takes place at the producer, not at the user.

6.4.6 Fines

This category includes any fines or penalties for lapses in regulatory compliance related to Waste and Emission Control. Several reporting guidelines request that these are disclosed separately, regardless of their amount in relation to other more significant expenditure. Also in the GRI guidelines indicator "EN 28" explicitly asks for: the monetary value of significant fines and total number of non-monetary sanctions for non-compliance with environmental laws and regulations.

The EC Recommendation and the Eurostat disclosure requirements define that costs incurred as a result of fines, or penalties for noncompliance with environmental regulation, and compensation to third parties as a result of loss or injury caused by past environmental pollution are excluded from the definition of environmental expenditure. Whilst related to the impact of the company's operations on the environment, these costs do not prevent, reduce or repair damage to the environment.

For internal EMA, these costs need to be recorded and monitored. For external reporting in sustainability and financial reports and to statistical agencies, they need to be disclosed separately.

6.4.7 *Insurance*

This category includes any costs of insurance covering potential liability related to Waste and Emission Control, such as insurance related to the accidental release of hazardous materials. Insurance covering higher risks of fire or other damage to the production site or at transport due to dealing with hazardous substances and dangerous processes could also be quoted.

The annual expenses for insurance are shown in the profit and loss accounts. But if insurance payments are required, companies must frequently foot part of the bill. Thus, even with risks covered by insurance, there may remain damage to be covered by the firm. This cost category is only relevant for specific industry sectors, mostly related to transport of hazardous goods or increased risks of some energy utilities.

6.4.8 *Remediation and Compensation*

This category includes any Remediation and Compensation costs related to cleaning up contaminated sites, recovery of contamination of land and water, compensation to third parties, etc. Examples of (contingent) liabilities which may emerge from company's activities include

- Groundwater contamination (e.g. from working with solvent-containing substances)
- Surface water contamination (e.g. from spills and transport damage)
- Air emissions (e.g. sudden release due to a break-down of pollution treatment equipment resulting in damage claims from neighbors)
- Energy emissions (e.g. radioactive emissions)
- Soil contamination (e.g. from contaminated surface water by missing protection troughs and collection tanks)

The liabilities for clean up and the necessity to account for them by annual provisions in the profit and loss accounts may primarily be derived from the rules of public law and, to some extent, civil and criminal law. Environment protection obligations under public law may include the duty to adapt equipment and procedures to the state of the art, to make provisions for waste removal and recycling at periodic intervals, to recultivate and dispose of substances at non-periodic intervals, and to clean up contaminated land.

6.4.8.1 Duty to Adapt Equipment and Procedures to the State of the Art

As a result of advances in the state of the art, industrial plants commissioned in the past may no longer meet the pertinent legal requirements. In order to comply with current emissions allowances, the law usually grants transition periods for existing plants that are liable for approval. While, from a legal point of view, the duty to adapt arises as soon as the applicable law takes effect, literature sometimes also stipulates an economic causal relationship in order for provisions to be formed.

6.4.8.2 Duty to Remove and Recycle Wastes

If there is a back-log, at the balance-sheet cut-off date, in compliance with mandatory waste removal and recycling duties arising at periodic intervals, this must be accounted for by the formation of provisions.

6.4.8.3 Remediation and Disposal Duties

Especially in mining or in connection with the erection and disposal of power stations and lines, there may be rules requiring comprehensive measures to restore the original landscape (e.g. run-of-the-river power stations) or controlled demolition of buildings (e.g. of nuclear power plants). This may also include compensation to farmers, fisheries and forestry.

6.4.8.4 Clean-Up of Contaminated Sites

Provisions for the clean-up of contaminated land may need to be taken when there is a likelihood of that duty arising, however, at the latest when the authority has knowledge of the contaminated site. In many countries, national tax laws require that a provision for future costs is calculated only once the legal obligation for this action has been established.

In general, whenever a company is required to repair a damage to the environment which has already occurred, especially in the context of cleaning up contaminated land, a provision because of the economic causal relationship in the past is possible, whereas the duty to adapt to new technical standards usually precludes provisions in view of future revenue, unless the duty to adapt already existed at the cut-off date. The EMA assessment should mirror the treatment the profit and loss accounts regarding annual postings of provisions.

6.5 Costs for Prevention and Other Environmental Management Costs

This category covers the costs of preventive integrated cleaner technologies as well as general environmental management activities. The main focus of this cost category is on annual costs for prevention of waste and emissions, but without calculated cost savings. This may include higher pro-rata costs for environment-friendly auxiliary and operating materials, low-emission process technologies and the development of environmentally benign products.

Prevention comprises for instance proactive eco-system management, on-site recycling, cleaner production, green purchasing, supply chain environmental management, ecodesign and extended producer responsibility. It also includes costs for general environmental management activities such as installing and maintaining an environmental management system, environmental accounting; environmental measurement (monitoring, performance auditing, performance evaluation, external certification); environmental communication (performance reporting, community group meetings, government lobbying) and any other relevant activities (such as financial support of environmental projects in the community).

This category includes costs for

- Equipment Depreciation
- Operating Materials, Water and Energy
- Internal Personnel
- External Services and
- Other Costs

Preventive activities such as on-site recycling, cleaner production and the implementation of an environmental management system have a special role to play for environmental protection. Costs incurred for preventive environmental management activities often not only improve environmental performance, but also bring a financial payback as materials and energy efficiency rises and waste declines. Accordingly, several technologies and projects are implemented not only to meet environmental targets, but also with efficiency, product quality or other goals in mind. The share of environmental protection for these costs thus needs to be estimated.

6.5.1 Equipment Depreciation

By definition, most equipment with effective pollution prevention is closely integrated into production equipment (such as cascading and closed loop water circulation systems that are an integral and automated part of some chemical manufacturing process). Often, the new technology also uses less energy, is faster and has more production capacity. A new bottling plant, for instance, is less noisy, requires less water, and is equipped with an automatic supply of detergents. Here, environmental protection is an inherent part of equipment design. In other cases, equipment (for example, a high efficiency paint spray gun) may simply contribute to Preventive Environmental Management because it inherently uses energy or raw materials more efficiently and produces less waste than alternative equipment. In such cases, an organization may wish to estimate what percentage (if any) of the operating costs for the equipment should be designated as "environment-related."

These estimates can be based on considerations for the primary reasons for purchase of that particular piece of equipment, for example, environmental or materials efficiency and on considerations regarding the actual environmental impact reduction.

It is recommended to estimate the percentage shares for the most relevant cleaner technologies on a cost center basis. Experience from the case studies shows that only for the most important technologies the total or partial costs from the cost center reports are being quoted in the annual EMA assessment template, while for other investments it was considered sufficient, that the material inputs are being quoted under the cost category for NPO.

If a specific cleaner technology was significantly more expensive because of integrated pollution prevention than other state of the art equipment with identical production values, and the investment was partly motivated by environmental considerations, than the related percentage may be quoted as an environmental investment and the annual operating costs may be recorded. However, if the cleaner technology represents the current state of the art and was installed mainly as a regular replacement of an old device, it should not be quoted as environmental investment.

Even if cleaner technologies may not show up under the cost category of depreciation, they significantly effect the environmental cost distribution over the years, as they contribute to reducing the amounts of operating materials, water and energy needed for production as well as personal required for environmental management and disposal costs.

6.5.2 *Operating Materials, Water and Energy*

As stated previously, Operating Materials are Materials Inputs that were never intended to leave the organization in the form of a product but are still necessary to run the organization. For equipment defined as integrated technology with a certain percentage of environmental relevance and posted on separate cost centers (see Section 6.5.1) the operating materials quoted there may be considered in this cost category but must be deducted from the cost category of NPO in order to avoid double counting.

6.5.3 *Internal Personnel*

Personnel dealing with prevention and environmental management should be recorded in this cost category. Again, some of the data may be taken directly from the cost center reports of the related equipment defined in Section 6.5.1. If a cost center for environmental management is installed, it can be assumed that all personnel costs collected their can simply be attributed to this cost category. But in addition it may be necessary to perform a screening of significant environmental management related projects and activities of the previous business year and estimate the time of the people involved. Such a screening will also be helpful to assess the costs of related external services.

The average person hours for people attending environmental trainings and participating in environmental protection oriented projects should be calculated based on average personal costs for this qualification level.

6.5.4 *External Services*

All external services related to Prevention and Other Environmental Management for consultants, training, contractors, inspections, audits, certification and communication should be included here. Also the costs for printing an environmental report and other communication-related activities like eco-sponsoring may be included. The related expenses will probably not have been systematically collected on one account or cost center but spread throughout the company and across accounts. A quick memory session on last year's projects and activities in the environmental team will make sure that all relevant expenditure can be traced back, and the allocation to expenditure items and cost centers can be improved. Often, the installation of a cost center for environmental management is the solution taken to insure consistent and complete recording of related services.

6.5.5 *Other Costs*

Any other relevant Prevention and Other Environmental Management Costs should be included here. Examples might be donations to environmental initiatives or nature reserves e.g. as an offset for CO^2 emissions caused from aircraft flights. While donations to environmental initiatives and nature reserves may be part of an organization's corporate social responsibility policy, they may also be used as compensation for environmental impacts in countries where environmental regulations are not as strict.

Examples for prevention related fees are

- Fees to register under environmental labeling schemes
- Fees for certification to environmental standards

Costs for environmental communication, e.g. for the publication of an environmental report or for environmental trainings may also be posted here.

6.6 Research and Development Costs

This category includes the costs of Research and Development activities on environment-related issues. Examples are research on the substitution of potentially toxic materials, application of recycled or renewable materials, development of energy-efficient products and testing of new equipment designs with higher material and energy efficiency.

Research and Development costs related to the environment might include costs of all cost categories, such as equipment depreciation, operating materials, water and energy, internal personnel and external services. Due to statistical reporting requirements they need to be reported separately. In many organizations, Research and Development is a separate department with its own cost center. Although research-related costs can be identified there, an organization may wish to determine which Research and Development costs is actually environment related and which are not.

6.7 Environmental Earnings and Savings

Environment-related Earnings may be gained from sales of by-products, sales of excess capacity of waste treatment facilities, revenues from insurance reimbursements for environment-related claims, subsidies for environment related research projects, investment grants for environment related equipment, etc.

On the contrary, Savings are realized only when a current, defined system changes in some way. For example, if efficiency improvements reduce materials and energy use and waste generation, the resulting monetary savings can be calculated by comparing the reduced costs to the previous, higher costs. These types of savings tend to occur when preventive environmental management activities are implemented. In order to be able to calculate savings, the costs of the previous business year or existing production equipment need to be available. EMA is the tool developed therefore.

The case study of the brewery Murau in Chapter 8 as well as Chapter 7.8 on calculating investment options provide further information.

6.8 Case Study of SCA Laakirchen Pulp and Paper Plant

SCA Graphic Laakirchen AG, one of SCA's pulp and paper production sites in Austria, has been tracking its physical and monetary information under EMA since 1999 and has a well-established, consistent system for capturing and assessing materials flows and environment-related costs. The information collected is used for decisions related to both environmental management and general production. SCA Laakirchen annually calculates total environment-related costs and disclosed their percentage distribution by environmental domain in its Environmental Statement (www.sca.at), as illustrated in Table 6.3.

The history of paper manufacture in Laakirchen dates back to 1874. The Swedish group "Svenska Cellulosa Aktiebolaget" (SCA) is structured into three business segments: "SCA Hygiene Products", "SCA Packaging" and "SCA Forest Products". Within the "Forest Products" group, the Laakirchen factory (SCA

Graphic Laakirchen) specializes in the production and development of super-calendared (SC) paper. SCA Graphic Laakirchen employs about 550 people and operates two paper machines–PM 3 and PM 10–which together annually produce around 330,000 tons of SC natural rotogravure and offset paper. SC paper from Laakirchen is used for magazines, catalogues and advertising materials with around 95% of production being exported, of which 80% goes to EU countries.

In addition to the demands of high quality in terms of printing, the principles of SCA Graphic Laakirchen also cover the need for environmental responsibility. Since 1993 the company has used totally chlorine-free bleached pulp (TCF) exclusively and has consequently adopted a leading role in the production of SC papers. The company was one of the first Austrian EMAS sites (registration number 23) and often is involved in pilot projects which are then implemented throughout the rest of the corporation. In 2000, the company conducted a pilot project on EMA, following the UN DSD approach, and has published in its annual environmental statement for the year 2000 both the report on the project and its environmental cost distribution over different environmental media (www.sca.at). For the research report, the following case study, which is based on the disclosure in the environmental statements with slightly modified and additional information was developed (Jasch and Schnitzer, 2002).

In addition to the equipment used directly for the actual paper manufacturing process, such as wood storage, grinding, stock preparation and the paper machines PM 3 and PM 10, SCA Graphic Laakirchen AG also has a variety of technical units which supplement environmental management. These are the de-inking unit for preparing recovered paper, a multi-level mechanical/biological wastewater treatment plant, and a gas turbine functioning in accordance with the principle of power/heat reaction, which guarantees a virtually self-sufficient energy supply for the plant thanks to its high level of efficiency. The annual depreciation of these units is recorded as 100% environment relevant equipment.

The total requirement for wood was around 240,000 m^3 per year, and 10 people worked in the wood storeroom. The materials loss of bark waste was 15% of the wood purchased. In manufacturing ground wood pulp, SCA Graphic Laakirchen used approximately 90% wood from thinned trees, primarily from Austrian forests. The materials loss at ground wood pulp production was about 1%. The materials loss percentages as well as the wasted work hours are recorded in the annual EMA cost assessment.

SCA Graphic Laakirchen used 130,000 tons of pre-graded recovered paper each year. About 80% of recovered paper input can be reused for paper production; the rest ends up as waste. Thus, 20% of the purchase counted as NPO. Fillers and totally chlorine-free bleached pulp (TCF) are bought in and dissolved on site as slurry. Production indicators calculated a loss of 0.4% of pulp and 4% for fillers.

After passing through the production process, all paper chemicals end up in the wastewater treatment plant and are thus recorded as NPO costs. The company ran a research project for a closed-loop system of paper chemicals, which was co-funded by the Austrian research fund. The profitability of this system is significantly higher if not only end-of-pipe-treatment costs are calculated, but also the savings made on purchases of materials are considered. For lubricants, it can be estimated that these end up together with the tissue, in the hazardous waste fraction. For cleaning materials, it was estimated that one-half leaves the company via air emissions, the other half via wastewater.

SCA Graphic Laakirchen AG operates a gas and steam plant in co-operation with an electricity generator. This is used for low-emission generation of electricity and steam from natural gas, and it ensures a largely autonomous energy supply according to the principle of power/heat reaction. It produced about 1,460 GWh of electricity and about 1,400 GWh of steam, and the efficiency loss of natural gas conversion was estimated at 30%. This cost center employed two blue-collar and two white-collar workers, 30% of their costs have also been calculated for the EMA assessment. There are no reliable estimates for the efficiency loss of electricity, so it was decided to calculate only the electricity inputs of environment-related cost centers.

The grinding shop had 14 continuous grinders for grinding the debarked wood. Work is continually being carried out to optimize the energy which is required for the grinding process. In subsequent stages, the wood pulp is graded and stored for further processing. The locations used for wood manipulation (wood transfer, the storeroom and the grinding station) were re-organized in 2000 in order to prepare for a ground-sealing. The costs which are expected for re-cultivation have been accounted for by a provision already in the previous balance sheet.

In addition to the environmental manager and the personnel costs of the specified cost centers, one technician per year, on average, is working on environment-related issues. In addition to the environmental manager, the company has an environmental board of seven people, which spends about 2 weeks per year on environmental issues. The director of the management board is also involved in environmental issues for about 1 week per year. All these costs have been calculated not on the basis of the actual salaries, but with average total personal costs for the different personal qualification levels.

The input output balance shows that not for all inputs and outputs volumes and monetary values were disclosed. It is based on the disclosure in the environmental statement for the year 2000 with slightly modified and additional information from the research project (Jasch and Schnitzer,2002) (Table 6.2).

Table 6.3 shows the percent distribution of total annual environmental costs for 2000. The language of the company's Environmental Statement has been modified to better match the EMA cost categories. As well, data subtotals were created. The rows show the costs by cost categories. The environmental statement reports that total environmental costs were 30% above the costs of the previous year. This substantial increase was attributable to increased prices for raw materials, operating materials and gas.

The data in Table 6.3 illustrate the fact that, in many companies, the "Materials Purchase and Processing Costs of NPOs" are often significantly higher than more familiar environment-related costs of "Waste and Emissions Control"–approximately four times as high in the case of SCA Laakirchen. Table 6.3 also illustrates the fact that the costs for "Prevention and other Environmental Management Costs" at SCA Laakirchen are quite low, despite the fact that the company has implemented a number

Table 6.2 Average input-output balance of SCA Laakirchen

	Unit	Volumes	€
Raw materials			
Wood	Tons	123,000	11,747,000
Pulp			26,105,000
Recovered paper	Tons	126,000	16,338,000
Packaging materials			2,264,000
Auxiliary materials	Tons	107,000	12,210,000
Operating materials			
Chemicals	Tons	16,000	8,137,700
Lubricants			109,100
Cleaning materials			32,700
Water	$1{,}000\,m^3$	6,725	11,600
Energy provision			
In-house hydro power	MWh	16,885	
Electricity: external	MWh	38,575	
Natural gas	$1{,}000\,m^3$	110,970	17,920,000
Energy consumption			
Electrical energy	MWh	460,509	
Thermal energy	MWh	400,003	
Production			
Graphic paper	Tons	323,000	
Waste water			
Waste water flow	$1{,}000\,m^3$	5,706	
COD	Tons	453.5	
Suspended solids	Tons	28.5	
Phosphor	Tons	2.1	
Nitrogen	Tons	7.7	
Air Emissions			
NO_x	Tons	106	
CO_2	Tons	221,000	
CO	Tons	55	
Waste			
Bark	Tons	3,679,000	
Fibre residues	Tons	909,000	
Flotation sludge	Tons	4,338,000	
Rejects	Tons	176,000	
Waste for recycling	Tons	5,000	
Waste to landfill	Tons	66,000	
Hazardous waste	Tons	5,000	

of preventive projects in past years. They have achieved significant savings in the cost categories for "Materials Costs of NPO" and "Waste and Emission Control."

The distribution by environmental domain shows that water/waste water was responsible for 54% of all environmental costs. The annual operating costs of the wastewater treatment plant accounted for 9.3%, but the purchase volume of the paper

Table 6.3 Environment related total annual costs at SCA Laakirchen–percentage distribution

Environmental domain environment-related cost categories	Air + climate (%)	Waste-water (%)	Waste (%)	Soil + ground water	Others	Total (%)
Materials purchase costs of NPOs						
Raw materials			15.2			15.2
Packaging			0.1			0.1
Auxiliary materials			2.7			2.7
Operating materials	0.1	42.2	0.5			42.8
Energy	19.8					19.8
Water		0.0				0.0
Materials processing costs of NPOs		0.2	1.0			1.2
Subtotal	19.9	42.4	19.5			81.8
Waste & emission control costs						
Equipment depreciation	0.1	2.8	0.4			3.3
Operating materials and services	0.2	5.5		0.1		5.8
Internal personnel	0.7	1.0	0.1			1.8
Fees, taxes and fines	0.9	2.7	6.0			9.6
Subtotal	1.9	12.0	6.5	0.1		20.5
Prevention and other environmental management costs						
External services fort env. management					0.4	0.4
Internal personnel for env. protection	0.1				0.2	0.3
Internal personnel for research & development					0.1	0.1
Subtotal	0.1				0.7	0.8
Environment-related cost total	21.9	54.4	26.0	0.1	0.7	103.1
Environment-related earnings total			–3.1			–3.1
Total Environment-related costs & earnings	21.9	54.4	22.9	0.1	0.7	100.0

chemicals which it disposes of down the drain is worth 42%. A reduction in the consumption of paper chemicals would therefore significantly reduce environmental costs. The column for Waste accounted for 23% of all environmental costs. Disposal fees were only 6%, but the major share was the purchase price of the raw materials which are included in the waste fraction (19.5%, including processing costs). The efficiency losses of gas combustion were assessed under the category of "air and climate". The sharp increase in the price for gas will probably raise the share of this cost category from 19.8% into 25% in the upcoming year.

Analysis of the environmental costs by cost categories makes evident that the highest share is the materials purchase value of non-product output (81.8%), which is calculated from all raw, auxiliary and operating materials in the mass balance that do not leave the company as part of the product. The earnings in the columns for "waste" and "wastewater" resulted from the sale of re-cycled materials and treatment capacity.

The research project on electrochemical wastewater treatment, which is also dealt with under "costs for research and development", was partly funded by the Austrian Fund for Research (FFF). These earnings are also accounted for; though more importantly, this project helped to reduce significantly the costs of paper chemical input.

The data in Table 6.3 allow SCA Laakirchen to compare its environment-related costs from year to year. Although manufacturing output rose almost 23% from the last business year, the use of a new paper machine kept the total environment-related costs increase to only 14.7% over the same period. This illustrates the overall positive financial impact of the company's environmental management initiatives. A more detailed look at the cost changes between years also revealed some interesting points. For example, the overall costs of operating the wastewater treatment plant did not change, even though it was enlarged to handle increased wastewater resulting from the expanded production. This was because the operational efficiency and maintenance of the wastewater treatment plant were improved in several ways as it was expanded.

Costs in other categories did increase. For example, the purchase costs of auxiliary materials increased not only because of expanded production, but also because of international price changes. Even though the distribution of total costs and earnings across the different environmental domains remained constant over the last years (22% air/climate; 54% wastewater; 23% waste; 1% other) the company expects sharp price increase for energy and thus a change in this distribution.

The physical results of SCA Laakirchen's environmental management efforts were also presented in the company's Environmental Statement. For example, despite the production increase of about 23%, the procurement of water increased by only 11%, the volume of wastewater by only 13%. These are increases in absolute terms, but are improvements per unit of production. Use of physical inputs, such as filler, recovered paper and energy, also increased in absolute terms but reflected eco-efficiency improvements.

Assessment of environmental costs following the UN DSD EMA approach has been carried out for several subsequent years, and has significantly changed awareness of the priority areas for cost savings. The focus changed from technical equipment and personnel hours to materials efficiency improvements.

Chapter 7
Linking Physical and Monetary Information

Chapter 7 focuses on linking the physical and monetary information system. It starts with consistency and consolidation issues to be considered when defining the system boundaries for an EMA assessment and when aggregating data from several sites or companies. The chapter deals with information available on the company level, traces environmental aspects in the balance sheet and where to find them in the profit and loss accounts. Section 7.4 goes one step further down into the organization and highlights the principles and terminology of cost accounting, process flow charts and overhead cost attribution. The concepts of activity based costing and material flow cost accounting are explained as well as where to get the necessary data from stock management and production planning systems. The last issues dealt with are application for investment appraisal, budgeting and benchmarking. Danisco, a global supplier to the food industry, uses EMA as a tool primarily to benchmark production sites, which are divers from a geographical and production process point of view in order to demonstrate differences and similarities.

7.1 Environmental Expenditure in the Profit and Loss Statement

The profit and loss statement may be arranged according to the expenditure or cost-categories-oriented format or to the operational (cost-of-sales) format. In the cost-categories-oriented format, all earnings and expenses of a period are listed. Operational expenditure is broken down into material and personnel expenditure, depreciation and other expenses. The accumulation and clearance of work in process and finished goods is determined by a stock-taking at year-end, assessed at production cost, and posted as correction of sales revenue.

In the cost-of-sales format, the actual sales of a period are compared only to those expenses which have been incurred for the manufacture of the products sold. The cost-of-sales format, therefore, requires a constant collection and assessment of inventory increases of finished products and work in process. The monthly earnings statement

C. Jasch, *Environmental and Material Flow Cost Accounting*: Principles and Procedures, Eco-Efficiency in Industry and Science 25,

thus leads to a more explicit operating result than the cost-categories-oriented format in which the changes in inventory are not recorded during the year although it is more sophisticated and time-consuming in terms of the cost accounting system used. The cost-of-sales format is structured differently and distinguishes between production costs of sales, and chronologically separates distribution costs, administrative costs and other operating expenses. The profit for the year is identical in both formats.

Tables 7.1 and 7.2 examine which accounts of the profit-and-loss accounts must be analyzed for EMA and under which cost categories of the environmental cost assessment scheme (as in Table 6.1) they are allocated. The cost-categories-oriented format is better suited for this purpose as the list of balances of the book-keeping department contains all the necessary information. In the cost-of-sales format, an analysis of both the accounts of the book-keeping department and of detailed cost center reports must be performed in order to determine total annual environmental costs.

Table 7.1 Cost-categories-oriented format of the profit and loss statement

Cost-categories-oriented format	To do
Turnover/net sales –Change in inventory	The output side of the material flow balance and the resulting distinction between product output and non product output are being assessed. Determine actual quantities produced, sales figures, losses on storage, spoilage, returns etc. Establish actual product output and loss of products between production and sales.
–Work performed and capitalized	May be relevant for production costs of in-house facilities for the removal, treatment and prevention of wastes and emissions (processing costs of NPO)
–Other operating income	Earnings from subsidies, investment grants and sales of non-product output
– Materials	Determine the material inputs by material categories and sub-categories. Determine share of non-product output of raw, auxiliary and operating materials and assess at material purchase costs. Energy and water supply costs should also be shown in this category, but are often posted under "other operating expenditure".
–Services (other external costs)	External services for maintenance of treatment facilities and cleaner technologies may be taken from the cost centres defined. All other services for general environment management, research and consultancy services, auditors, trainings, external information and communication etc. will be scattered across a variety of accounts.
–Personnel expenses	Record personnel costs from cost centres defined as environmentally relevant with the appropriate share. In addition, determine work hours of staff not traceable from cost centre reports, e.g. general environmental management activities, work spent on specific projects and for trainings. Multiply by average work hour rates as established by in ternal calculation procedures.

(continued)

Table 7.1 (continued)

Cost-categories-oriented format	To do
– Depreciation	Define waste and emission treatment equipment by cost centres. Define cleaner technologies and determine if they have been significantly more expensive in relation to state of the art or if they have a significant share of environmental impact reduction. If yes, define the percentage. Record investment costs, year of put in function and the related depreciation.
– Other operating expenses	Record expenses from the cost centres defined as environmentally relevant. Conduct a brain storming with the environmental team and production manager on significant projects, activities and other costs of last year to check for completeness of the costs recorded already. Transport expenditure for wastes, disposal and collection fees, licenses, printing costs for environmental reporting, registration fees, eco-sponsoring, penalties, insurance premiums, provisions etc. are scattered across a variety of accounts. The checklists included in the annex are designed to assist the user in tracing and assessing these costs. Also purchase costs of power, fuel and water can sometimes be found in this category, even though they belong under 'materials'.
– Other taxes	Environmental taxes, disposal and connection fees may be posted under this category but may also be mingled into several accounts of other operating expenses.
= Operating profit, EBIT earnings before interest and tax	
+/– Financing	Not relevant for EMA. If environmental costs are assessed instead of expenditure, pro rata financing cost for depreciation of fixed assets and production costs of NPO may be calculated.
= Profit (loss) on ordinary activities (after financial items and before tax)	
+/– extraordinary results	Normally not relevant for EMA, except in the case of break-downs and accidents and sudden discoveries of contaminated sites.
– Taxes on income and earnings	Not relevant for EMA.
= Net earnings/Profit after tax	

Organizations applying the cost-of-sales format typically have much advanced structures of posting costs to cost centers, but without the help of the cost accountants it becomes quite impossible to find anything which has been posted outside the sphere of influence of a cost center manager. Material and production-related direct costs and special direct costs of production are always shown under "production costs". Material-related direct costs include raw and auxiliary materials as well as packaging materials allocated directly to a product, depending on the cost

Table 7.2 Cost-of-sales format of the profit and loss statement

Cost-of-sales format	To do
Sales revenue	Relevant only, if production volumes are not available
–Production costs of goods and services supplied to achieve sales revenue	All costs relevant for EMA should be included here. Break-down according to accounts and cost centres
= Gross earnings from sales	
–Distribution costs	May include some less relevant costs, e.g. the costs for the environment report and other communication
–Administrative costs	Not relevant, unless containing environmentally relevant costs, not posted to production costs
Other operating revenue	Check for subsidies, investments grants and revenue from residual materials sold
–Other operating expenditure	Check for cost missing in the production costs of services
–Other taxes	Check for cost missing in the production costs of services
= Operating result	

accounting system used. Production-related direct costs comprise wages in production, allocated according to work hour records and cost centers. Material- and production-related overhead costs (other labor costs, operating materials, depreciation's for production plants) may be posted under production costs or under the item "other operating expenditure". To be able to determine the appropriate share of non-product output, a rather detailed break-down together with the corporate cost accountant must be performed.

7.2 Improving the Consistency of Materials Inputs and Product and Non-Product Output

Inputs and Outputs of materials can partly be derived from the profit and loss accounts, which should provide at least a complete record of all materials purchased in monetary terms. Systems of increasing complexity are being used for warehouse management, depending on the size of the company and the value of materials and products.

Ranked by complexity of the information system examples for recording of materials inputs are

- Material purchase value is recorded as expenditure directly at procurement; a further tracking of quantities used is not possible. This system is common in small companies, for the service sector and for operating materials.
- Material stock numbers are used to record material quantities as well, but materials are not monitored via storehouse management. This system enables

determination of annual quantities purchased, but not the point and time of consumption in the company.

- Materials are posted with material numbers to the incoming store. Inventories are being performed either annually or monthly to cross check for actual quantities used.
- When needed for production, materials are being called from the production planning system with an internal order form. For the materials included in this system their input into production can be determined exactly by value and quantity. Often, this system is applied only for raw and some auxiliary materials, but not for operating materials.
- Consumption of raw and auxiliary materials is posted to cost centers.
- Also all operating materials are posted to the incoming store and have to be called by internal orders, which allows posting of their consumption to cost centers.
- Waste and disposal costs and quantities are also recorded via storage management by way of internal records.
- Waste and disposal costs are in addition assigned to the relevant costs centers by means of records from the waste management collection team.

The following systems for recording of materials outputs may be differentiated:

- Only turnover is known, not the actual production volume; the losses on the outgoing store and for internal use are not recorded or only recorded in a total monetary figure at the end of the year for the inventory taking.
- There are production statistics.
- All materials produced are posted in the outgoing store and delivered to customers with a separated order form.
- The production planning system calculates estimated input and output based on the recipe. Planned consumption is cross checked with actual consumption by means of internal order forms.
- Product output and non-product output (scrap, losses, waste and emissions) can be tracked by cost centers.

Differences between material purchase and material consumption for production may be significant. Apart from the time lag, which may be costly from a financing point of view, losses on interim storage can cause considerable waste and costs that can be traced to a variety of causes. Losses are frequently caused by employees' private use of materials, material aging in the warehouse, becoming obsolete or unusable, or contaminated through careless treatment or otherwise destroyed.

In part, discrepancies between production output and sales volumes may have similar causes. In addition, there may be discrepancies due to internal usage within the company, returns, quality control, repackaging for different destinations or customer requirements, etc.

Discrepancy between materials consumed and production output reflect actual process-based waste, scrap and emissions. This comparison is distorted if material purchase must be compared against sales because of inadequate internal data systems. Inventory losses should be addressed separately as each type of loss requires different actions for improvement. The amount of sales (turnover) is only relevant for EMA if there are significant differences to production volume and investigation of the underlying causes or, if no better data is available for production volumes.

Ideally, the comparison of materials Inputs and product Outputs should be able to separately record the losses on the incoming and out coming stores from the losses during the production processes, as the measures to be taken to reduce these losses are quite different. However, this is only possible if a production planning system is installed, which separately records raw, auxiliary, packaging and operating materials, as well as waste volumes. Production planning systems can deal with thousands of materials. Amounts are recorded when materials are ordered or stored, and again when they are taken out of storage and moved into the production process. But, experience shows that the data is not often nor cross checked for consistency with technical monitoring systems and with data requirements of technical monitoring systems.

Checking stock accounts and flow accounts for consistency is crucial for any material flow balance. At least for the different raw and auxiliary materials inventory management should allow separate recording of each material input with its related inventory stock volumes at the end of the business year. However, experience shows that often the data is only available by separate material numbers or in total aggregated format and only in monetary values. In many companies the volumes consumed by material group can thus not be calculated (see Fig. 7.1).

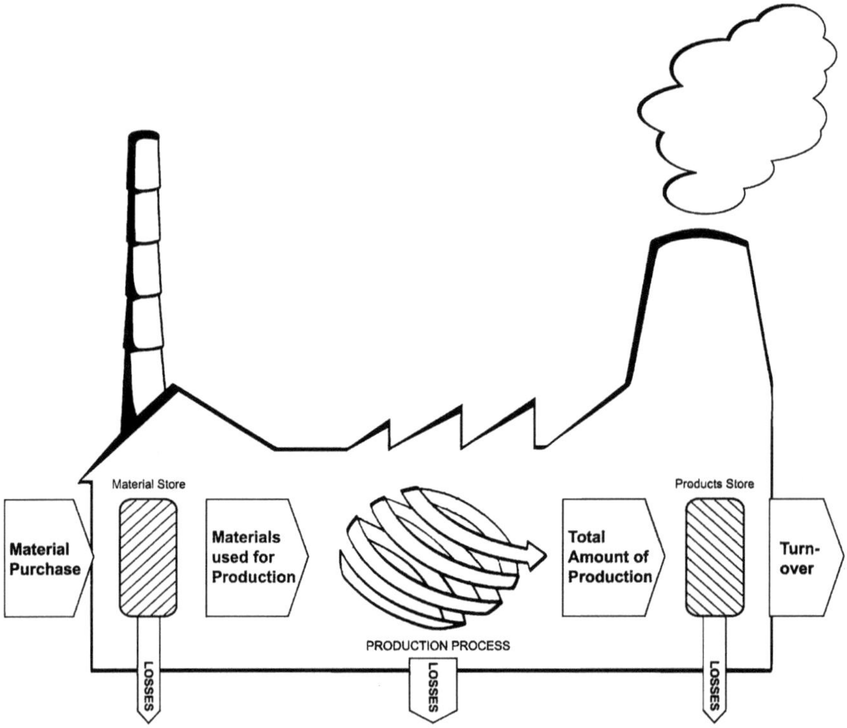

Fig. 7.1 Differences between material purchase and use for production

In order to be able to perform a consistency check between materials inputs and related outputs the data must be recorded in the same units, preferably volumes in kilograms. Recording of units of materials used (such as five boxes of paint) only makes sense if the production planning system has conversion factors installed that automatically correlate processed units to the resulting products. Actual monitoring and recalculation of estimated consumption ratios provides insight into saving potentials. All relevant information such as price, quantity, conversion factors and material numbers should be recorded at once from the supplier invoice.

The material flow balance can be checked for consistency by comparing it, to the extent possible, to material supply from stock keeping, sales information and production lists. For raw, auxiliary and packaging as well as final products, this may be easily done within the existing systems.

It becomes more complicated, however, if the majority of materials, and probably all operational materials, which are the ones often significantly impacting the environment, like chemicals, paints and lacquers, cleaning materials, workshop needs, etc. cannot be traced by material numbers. Often, all these materials vanish on stock and in overhead and the related values and volumes cannot be traced.

Several case studies showed that companies include only raw materials and some packaging materials in direct costs, but not auxiliary and operational materials, other packaging materials and the cost of disposal. Therefore, the consistency check provided significant potential for improved classification of accounts, the logic of assignment of material numbers and aggregation possibilities and the posting of material consumption to cost centers.

In the interest of efficient use of information (and in order to eliminate the need to go back to original invoices as sources of information) it is recommended that the departments involved define a procedure for gradually improving the recording of materials by material numbers and on stock inventory. Purchase and Material and Warehouse Management thus have an important role to play in EMA regarding the system design for the input output material flow balance.

A procedure of such a cross check of materials purchased, production output, conversion factors and the recipes applied by the production planning system, is shown in Figs. 7.2 and 7.3. Experience from case studies shows that often scrap percentages, which have been rough estimates, need adjustment. Automated cutting and dosage plants frequently have much better amortization times than expected since actual losses are often higher than estimated. Several companies in the production sector established a monthly reporting system where the cost center report and materials flow report is automatically generated from defined data monitoring systems and cross checked with additional measuring at least annually at inventory taking.

For the first round of an EMA assessment of the previous business year it is sufficient to account for about 70% of all material input in values and estimate the related amounts in volumes, if not available. Likely results of the EMA assessment could be

- Adjustment of the percentages used to calculate scrap resulting from raw and auxiliary materials, packaging and products
- Improved monitoring of materials and products in stock by material numbers

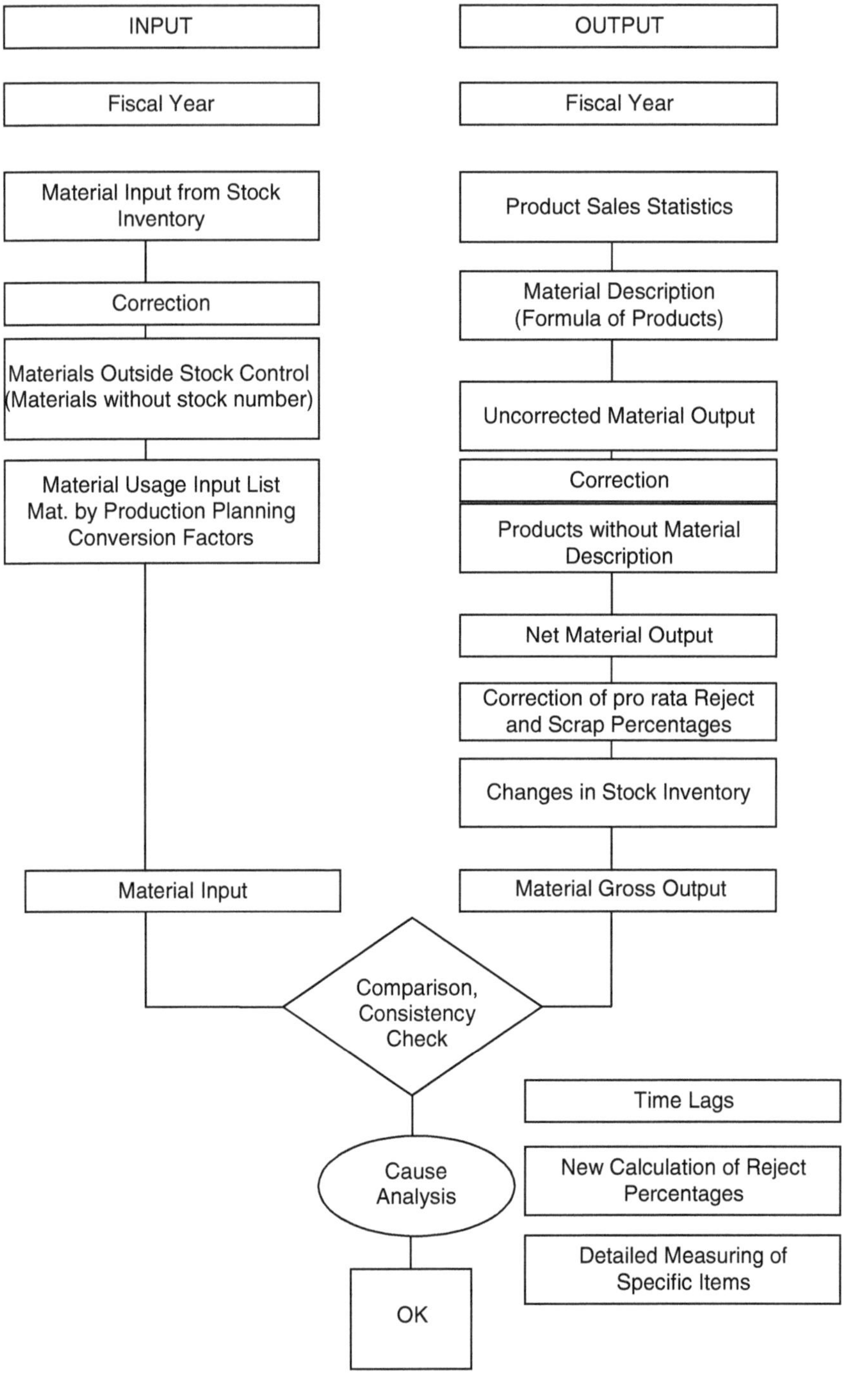

Fig. 7.2 Consistency check with the production planning system

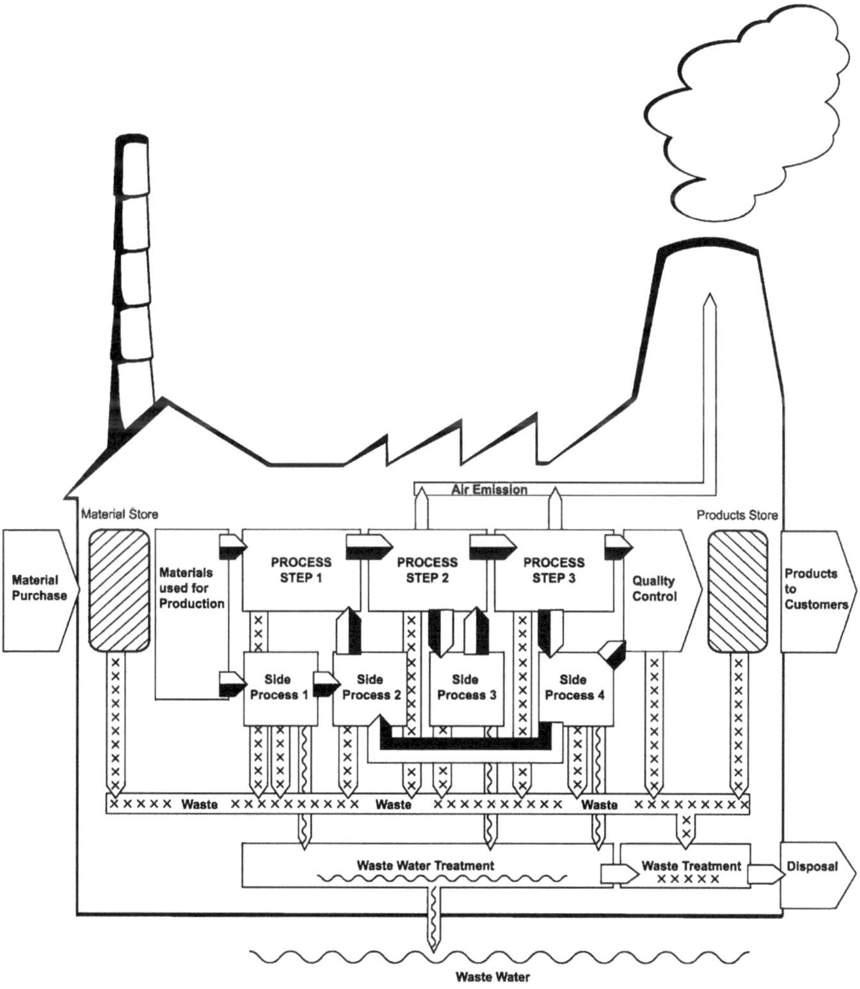

Fig. 7.3 Mapping material flows in financial and technical accounting systems

- Installation of new accounts and/or cost centers
- Defining a procedure for gradually integrating all materials input into production planning and stock management
- Regular monitoring of scrap and waste volumes installed at quality control, preferably by cost center
- Improved consistency of information systems and records behind with defined interfaces
- Depending on the business sector, installation of computer-aided design and cutting machinery
- Automated dosage equipment for operating materials and

- A marked improvement and consistency in information systems and records based on them

In some business sectors there can be significant time lags between material purchase, material use in production, the finished product being put on stock and final delivery to and invoicing of the customer. As production patterns change, emissions may occur much later than material inputs or product output. These time lag can be minimized once the material flow balance relates material input into production (consumption and not purchase) to relating product output of production (and not product turnover).

But, sometimes, the hindrances to overcome are not only related with time and money. Caution is needed as information is also a source of power in organizations. Departments, which have been used to purchase out of their own budget may also not be interested to change for a system that requests that all materials are being order via the warehouse management with defined material numbers.

7.3 Tracing Materials in Corporate Information Systems

The tracing of physical information on the flow of energy, water, materials and wastes is important under EMA because such information allows an organization to assess (and report) the important materials-related aspects of its environmental performance. In addition, materials purchase costs are key cost drivers in many organizations.

Much of the required physical accounting information unfortunately is not easily available to accounting personnel, as it is not systematically recorded or is not recorded in a way that reflects the real-world flow of materials. Personnel in other areas, such as production, environment or cleaning and waste management, generally have more detailed estimates and measurements of physical flows of materials, but often this information is not cross-checked with that of the accounting department.

Under the physical accounting side of EMA, an organization should try to track all physical inputs and outputs and ensure that no significant amounts of energy, water or other materials are unaccounted for. As this terminology implies, the underlying assumption is that all physical inputs must eventually become outputs—either physical products or waste and emissions—and the inputs and outputs must balance. The level of precision of a materials balance can vary, depending on the specific purposes of the information collection and the availability and quality of the data.

For a complete and integrated picture of materials use, the details of materials flows must be traced through all the different organizational materials management steps, such as materials procurement, delivery, inventory, internal distribution, use and product shipping, as well as waste collection, recycling, treatment and disposal. This can best be achieved if materials are assigned materials numbers.

In order to compile a material flow balance, it is recommended to apply a top down approach and start with the lists of accounts of conventional bookkeeping. Only this list provides a complete overview (in monetary terms) of purchased raw materials, auxiliary and operating materials in a given month or year. Each account of the profit and loss statement should be examined to determine whether materials may have been posted there. Personnel costs are not considered in a material flow balance.

The next step is to break down the accounts and specify the material groups in more detail. This might result in a recommendation for establishing additional accounts for the recording materials inputs.

Based on the input-output scheme in Tables 1.1 (Executive Summary) and 2.1 and the material groups recorded so far, a first breakdown for the I/O can be done. Next, information sources need to be identified and the recording of material groups discussed and probably improved. The tracing matrix for material flow data in Table 7.3 provides an overview on how materials are being recorded and where improvements may be needed.

The first column shows the structure of the material flow balance, which should be further detailed according to the company's needs. Table 6.2 provides an example for a pulp and paper plant, Table 8.1 for a brewery. The matrix serves to examine data consistency and the relationship between the material flow balance and existing information systems and documentation. The first round of mass balancing will discover inconsistencies and information gaps, which will enable improvements in the organization of internal data.

It should be determined:

- In which unit (Kilogram, Liters, m^3, or pieces) is the material recorded?
- What is the purchase value in a given year?
- What amount and value has actually used for production?
- On what account is the material posted?
- Is the material recorded with material number (this normally enables the direct recording of volumes together with values, even if the material is not monitored via warehouse management)?
- Is the material included in warehouse management and stock keeping?
- Is the material included in the recipes (formulae) of the production planning system?
- Is the material regarded as direct cost or overhead in cost accounting?
- To which cost center(s) is the materials use assigned to?
- Are there additional records or measurements, e.g. scrap production reports, waste collection reports by cost center?
- Is the data measured or estimated?

The tracing matrix for material flow data has been designed to enable an overview of how materials are currently being recorded in the information systems and where to start with improvements and closing of information gaps. At the same time, it also serves to correlate data (through data processing).

It is important to define quantity units as uniformly as possible and to give preference to kilograms. It doesn't make sense to determine the units of materials used because they cannot be correlated with the output side.

Table 7.3 Tracing matrix for material flow data

Tracing matrix for material flow data	Unit for the mass balance (kg, L, m^3, kWh)	Purchase value	Materials consumed value	Materials consumed volume	Account number	Material stock number	Stock keeping	Production planning syst.	Direct costs	Overhead	Assigned to cost centre	Oth.	Calculation/ estimates
Raw materials	√	√			√	√	√	√	√		√	√	
Auxiliary materials	√	√			√	√	√	√	√		√	√	
Packaging	√	√			√	√	√	√	√		√	√	
Operating materials	√	√			√	√				√	√	√	
Energy	√	√			√					√	√	√	
Water	√	√			√					√		√	
Product output	√	√			√							√	
Waste	√				√					√	√	√	
Waste Water	√				√					√		√	√
Air-emissions	√									√		√	√

Such an assessment may suggest

- The creation of additional accounts
- The generation of additional material stock numbers
- Categorization of which material numbers are collected on which accounts
- Assignment of certain material groups (e.g. operating materials) to warehousing or production planning systems (e.g. packaging material)
- Reorganization of cost accounting and
- Creation of additional records, especially with regard to emissions

Clear definitions as to which elements of the Input/Output analysis are recorded in which accounts, which material numbers are assigned to which accounts and which materials are also recorded in stock management are essential. A □ indicates the availability of data in the information system referenced. The objective should be to gradually improve the recording of material flows on a step by step procedure. There is not point in being complete in the first year; the goal is to gradually trace materials as completely and consistently as possible, in storage administration, cost centers and in production planning.

7.4 Cost Accounting Basics and Terminology

There is a continuous exchange of data and information between financial accounting, cost accounting, budgeting and controlling. Aside from this information and data exchange, cost accounting has the following main objectives:

- Identification of price floors and ceilings
- Calculation of planned and past production costs
- Evaluation of internal services, finished and unfinished products for sales or tax purposes
- Improving economic efficiency
- Providing data for company policy and decision-making
- Short-term performance evaluation
- Benchmarking of cost centers and production sites

Cost accounting is clearly distinguished from financial accounting by its calculating procedures. Its primary objectives are cost controlling, monitoring and planning (Table 7.4). However, many companies, especially small and medium sized companies (SMEs), work with data from the profit and loss accounts. It is up to management to decide whether the company should use cost accounting, and if so, which system it should use and how it should be designed. In contrast to financial accounting, this decision is not influenced by tax and commercial law.

The following definitions apply

Fixed Costs are not related to production volume, such as rent, interest on bank loans etc.

Variable Costs are directly related to production volume, e.g. raw materials and production labor hours.

Table 7.4 Relationship between cost category, cost center and cost carrier accounting (Adopted from Jasch et al., 1997)

Cost category accounting		Cost centre accounting		Cost carrier accounting (product)
Which costs have been incurred in which amounts?		Where and in which amounts have which costs been incurred during the accounting period?		Which types of costs have been incurred in which amounts for a certain product or service?
Cost distribution to direct costs and overhead				
Cost roll-over from financial accounting		Internal cost attribution and cost estimates or billing rates		
e.g.		e.g.		
Depreciation	→	I Manufacturing in several production steps	→	Product A
Raw and auxiliary materials				Product B
				Product C
Operating materials		II Warehouse		
Energy		III Distribution		
Internal personnel		IV Waste water		
External services maintenance		Treatment		
		V. Environmental management		
Other operating costs				
Calculated interest		VI. Administration		
Calculated risk				
↓	→		↑	

Direct Costs are attributed to the corresponding cost centers (process steps) and cost carriers (products). They include at least raw materials and production wages.

Overhead Costs are costs not attributed to cost centers by invoice or other means of recording, but collected in overhead and assigned to cost centers on an average basis. There are a number of methods to attribute overhead to cost centers and cost carriers.

Calculated Costs are used in cost accounting because they are not—or at different values—available from financial accounting, but influence operating results. If these costs are not matched by expenditure in financial accounting, they are called extraordinary costs, e.g. calculated equity capital interest, and calculated rent/lease, calculated management wages. If these costs are matched by expenditure in bookkeeping, they are also called Other Costs such as calculated borrowed capital interest, calculated write-offs on the basis of replacement prices, calculated risks; etc.

Costs Centers are those parts of the company that are organized as independent clearinghouses; they should be connected to production processes. Maximum consistency between cost centers and process-oriented material flow analyses is the prerequisite for good data for MFCA. Cost centers generate costs, are responsible for costs or are being attributed costs, e.g. for production and administration.

Cost Carriers or Objects are products and services produced either for the market or for internal needs. By attributing costs to cost centers and cost carriers, production costs and sales price floors are being calculated.

Cost accounting is performed in several steps. First, the costs are collected on cost centers related to production steps and additional cost centers like the waste water treatment plan or administrative processes. In the next step the cost of the additional cost centers are levied back to the production cost centers. This can be done as general overhead or on a more detailed and process or product specific basis. Lastly, the costs form the production cost centers are attributed to the respective cost carriers (i.e. products A and B).

Cost-Category Accounting is the first step in cost accounting and answers the question:
Which costs have been incurred in which amounts during the accounting period?

In cost-category accounting, costs are recorded in comparison to budgeted costs and divided into direct costs and overhead. It may require a roll over from financial accounting to cost accounting, as calculatory values may be used for cost accounting.

Cost Center Accounting is the next step and answers the question:
Where and in which amounts have which costs been incurred during the accounting period?

The overhead allocation sheet is used for internal cost assignments to cost centers. Finally, cost center accounting may determine billing rates (or surcharge rates) should they be required for cost carrier accounting based on the company operational situation.

Cost attribution is done in two steps, first from supportive (e.g. the environmental) cost centers like waste management and emission treatment, to the responsible cost centers in the production process and secondly from the production cost centers to the respective cost carriers/objects (product A and B).

Cost Carrier Accounting is the final step of cost accounting and determines the production costs for each product (or service). It provides the basis for price calculation. The question answered is:
Which types of costs have been incurred in which amounts for a certain product or service?

7.5 Mapping Costs Centers, Production Planning and Technical Monitoring

In the recommended top down approach for material flow cost accounting the next step after the environmental cost assessment and material flow balancing on a corporate level is to allocate the data from the system boundary of the company fence to internal processes, preferably by applying the structure of the cost centers defined.

Process flow charts, which trace the inputs and outputs of material flows (solid, liquid and volatile) on a technical process level, give insights into company-specific processes and allow the determination of losses, leakages and waste streams at the originating source. This requires a detailed examination of individual steps in production—again in the form of an input-output analysis, but sometimes linked to technical Sankey diagrams. The process flow charts combine technical information with cost accounting data. They are mostly not done on a yearly basis but for a specified production unit, machinery or cost center. In total, they should aggregate to the yearly amount.

This includes all material flows along the value-added chain, from incoming goods, via the various production processes, through product distribution to the customer. It also includes all the material losses incurred at various stages along the logistics chain (e.g. rejects, scraps, chippings, destruction of expired items or damaged goods), which leave the company as environmentally and economically undesirable non product output (solid waste, effluent, emissions). The corporate material flow balance is divided into various production steps and cost centers.

The process level is the main focus for pollution prevention activities. Data on the process level is also necessary for further analysis by products. This level of material flow analysis will be in the responsibility of technicians, but the data gathered should be cross-checked to ensure consistency with the cost accounting system. Usually a harmonization of technical data with data from accounting is not undertaken due to lack of inter-departmental communication. Experience has shown that such a consistency check provides great optimization potentials, and has thus become a major tool in environmental accounting. Therefore it is desirable that the technical and financial accounting have defined interfaces which allow cross checks of the data provided.

For greenhouse gas monitoring such a consistency check has become mandatory. As CO_2 emissions are not recorded based on the emission volume, but calculated based on the inputs of energy carriers and raw materials into specified processes, the monitoring regulations require a confirmation of the data on material and energy inputs from financial accounting.

The procedure recommended therefore is to visualize the structure of cost centers of the company involved like in Fig. 7.3 and mark define points of data gathering as in Fig. 7.4. In the next step, the structure of the technical monitoring system is visualized (Fig. 7.3) and again points for data gathering are highlighted (Fig. 7.4). The last step is to cross check the data provided by the two systems and to record and improve interfaces between the financial and technical information system.

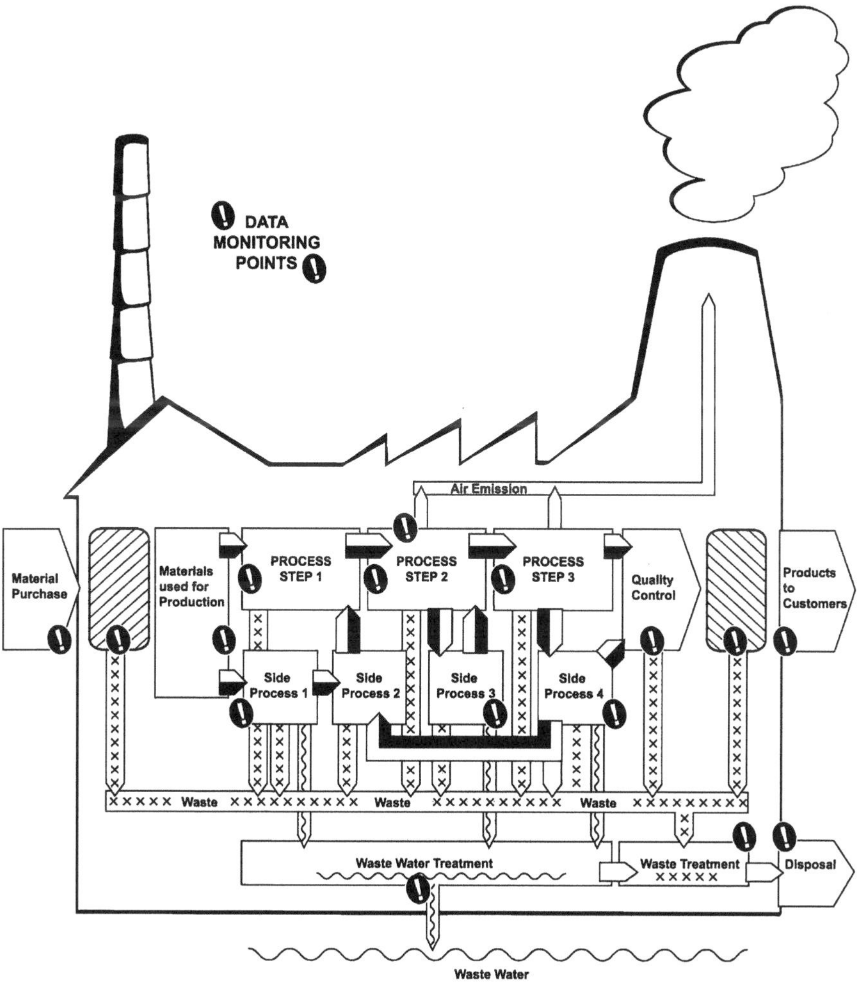

Fig. 7.4 Visualizing points for data monitoring

7.6 Activity Based Costing

The focus of activity based costing (ABC) is correct allocation of costs to processes and products, thus reducing the amount of costs hidden in overhead cost categories. Whenever possible, costs should be allocated to the respective cost centers and cost carriers/objects (products). Many terms are used to describe the methodologies for that purpose, such as "activity based costing", "full cost accounting", "process costing", and "material flow costing".

Activity based costing improves cost accounting by allocating costs typically found in overhead costs to the polluting activities and products. The strength of

ABC is that it enhances the understanding of the business processes associated with each product. It reveals where value is added and where value is destroyed. Significant material flows are traced throughout the company and their costs are allocated back to the polluting cost centers. Applying this approach can improve economic performance as a consequence of improved environmental protection. Moreover, ignoring this approach may result in incorrect product pricing and investment decisions.

The simple example in Tables 7.5 and 7.6 shows how overhead cost-attribution can significantly change the production costs of products. However, awareness is necessary, that changed allocation rules may imply a redistribution of power. Production lines and products which used to be profitable may suddenly have a bad performance, so the responsible line managers will tend to refuse the change, especially if the don't have the means to improve their situation. The example shows that costs of "joint" environmental cost centers, such as waste water treatment plants, but also operating material inputs like energy consumption, should be differentiated from other overhead costs and allocated to the related processes and products.

In the example a manufacturer has two products. Only materials and working hours are recorded as direct costs. All the other costs, including energy input and waste and emission treatment are treated as overhead and distributed to the products based on their turnover, which is assumed equal for both products. Table 7.5 resultantly calculated identical production costs for both products.

Internal environmental costs are often treated as overhead costs and divided equally between all cost drivers. A common example is that the costs of treating toxic waste of a product are included in the general overhead costs, and the overhead is

Table 7.5 Environmental costs hidden in overhead accounts

			Example		
	Product A	Product B	Overhead	Product A	Product B
Materials by recipe/ formula and stock issuing	Direct costs	Direct costs		70	70
Working hours by time records	Direct costs	Direct costs		30	30
Overhead	**Distribution by % product turnover**				
Depreciation			50		
Rent			10		
Energy			5		
Communication			10		
Administration			25		
Top management's salary			10		
Waste & emission treatment			10		
Total overhead			**120**	**60**	**60**
Total product costs				**160**	**160**

Table 7.6 Environmental costs attributed to cost centers and products

			Example		
	Product A	Product B	Overhead	Product A	Product B
Materials by recipe/formula and stock issuing	Direct costs	Direct costs		70	70
Working hours by time records	Direct costs	Direct costs		30	30
Energy	**Attribution to cost centres and products by actual process flows and equipment involved**		1	1	3
Waste and emission treatment			1	3	6
Depreciation			7	13	30
Overhead	**distribution by % product turnover**				
Rent			10		
Communication			10		
Administration			25		
Top management's salary			10		
Total overhead			**64**	**32**	**32**
Total product costs				**149**	**171**

allocated in equal parts to all products. However, 'dirty' products cause more emissions and require more clean-up facilities than 'clean' products. Equal allocation of those costs therefore subsidizes environmentally more harmful products. The clean products, on the other hand, are 'penalized' by this allocation rule as they bear costs that they did not cause.

The simple example in Table 7.6 illustrates how equal allocation can lead to suboptimal management decisions. In the example it is assumed that product A is 'clean' and does not cause any environment-driven costs for the company, while product B requires additional energy input and produced more waste and emissions, e.g. because product A is plain wood, while product B is lacquered and thus results in treatment requirements in the waste water treatment plant. If energy input, waste and emission treatment and the related depreciation are allocated to cost centers and products by actual process flows and equipment involved, the production costs of products are significantly changed. The "clean" product now only has production costs of €149, while the "polluting" product has production costs of €171.

Suboptimal cost allocation thus significant influences the pricing of products. The cross-subsidized dirty products are sold too cheaply whereas the environmentally less harmful products are sold too expensively. In consequence, market share is lost in more sustainable fields of activity.

Whenever possible, material flows and environment related costs should thus be allocated directly to the activity that causes the costs and to the respective cost centers and cost drivers. Consequently, the costs of treating, for example, the toxic waste arising from a product should directly and exclusively be allocated to that product.

The choice of an accurate allocation key is crucial for obtaining correct information for cost accounting. It is important that the chosen allocation key is closely linked with actual, environment-related costs. Turnover or production hours are thus not recommended. In practice, the following allocation keys are recommended for environment related issues:

- Volume of emissions or waste treated
- Relative costs of treating different kinds of waste or emissions
- Direct costs of material inputs, treatment or projects

7.7 Material Flow Cost Accounting (MFCA)

MFCA is a tool for measuring the flows and stocks of materials for a company, a production process or product in both physical and monetary units. It is based on an input-output analysis of material flows as described in Chapters 2 and 3, but applies a different cost allocation procedure. The German Federal Environmental Ministry and Federal Environmental Agency define: "MFCA regards the relevant material flows as cost collectors, and therefore allocates the costs of the company's production operations to these material flows" (2003).

MFCA strongly supports increases in energy and material efficiency (Fichter et al. 1999). In MFCA all input materials are traced and categorized as "product" or "non-product" (material loss), as explained in Chapter 2, 3, and 6. The products sold are called "positive products" while waste and emissions are called "negative products" or "non-product output".

In the IFAC and UN DSD EMA approach the NPO is calculated with its material purchase value. In addition, production costs for NPO may be calculated, which is mostly done for products, which have been claimed as below quality and which therefore also end up in waste. The focus of MFCA, in contrast, is to allocate all production costs to material flows. On the other hand, MFCA does not calculate environmental costs or waste and emission treatment or integrated prevention and environmental management. It focuses on the costs for product and non product output.

ISO TC 207 Environmental Management as adopted a new work item on MFCA in March 2008. Annex B of the new work item proposal (ISO/TC 207 SC N 856, 2008) provides a simple example highlighting the different costs calculation procedure between conventional cost accounting and MFCA. The production process in the example produces one product from 100 kg materials with materials purchase

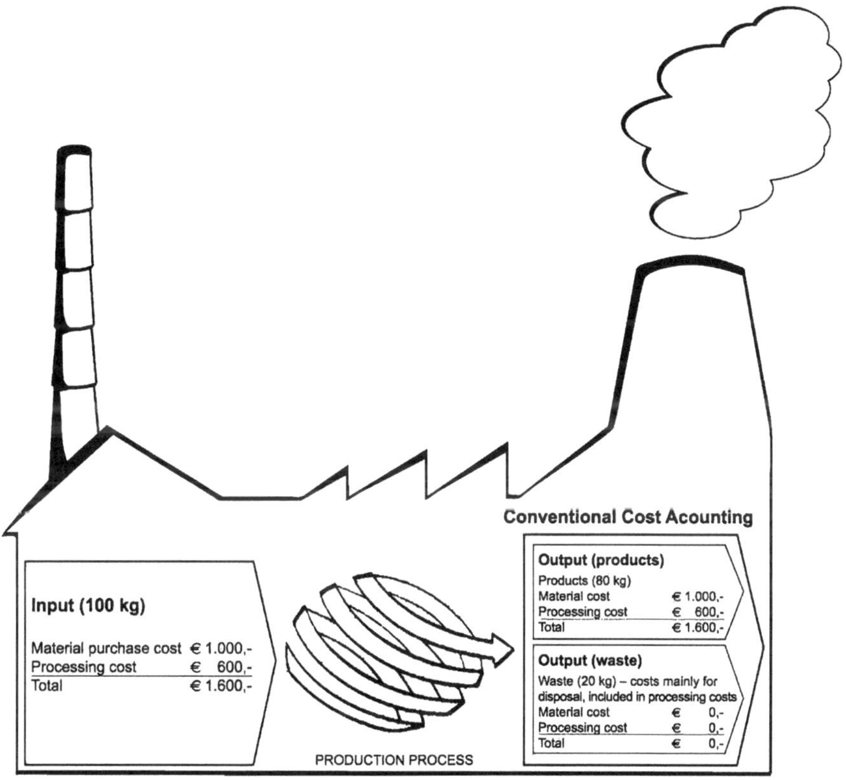

Fig. 7.5 Cost of waste in conventional cost accounting

costs of €1,000 an processing cost of €600. As Fig. 7.5 (adapted from ISO/TC 207 SC N 856, 2007) shows, the value of the waste is not recorded separately in conventional cost accounting, but automatically included in the cost of product output. Mostly, companies include disposal fees in general overhead which is levied to production costs.

However, in MFCA waste is treated as a separate negative product, and proper amounts of costs are allocated to it based on the weights of product and non product output. Therefore, the value of the waste as a negative product is calculated as €320 in Fig. 7.6. This information provides an incentive to management to reduce these costs. In well developed MFCA systems this cost allocation is performed for each production step, which results in considerably high costs of negative products.

MFCA is a tool especially useful for manufacturing companies, but has also been applied in other sectors including primary and service industries. The tool is useful to all organizations which handle or consume materials including energy. In addition, MFCA may also be extended beyond the system boundary of a single company to realize resource productivity improvements across the entire supply chain. As with the input-output balance described in Chapter 2 & 3, material

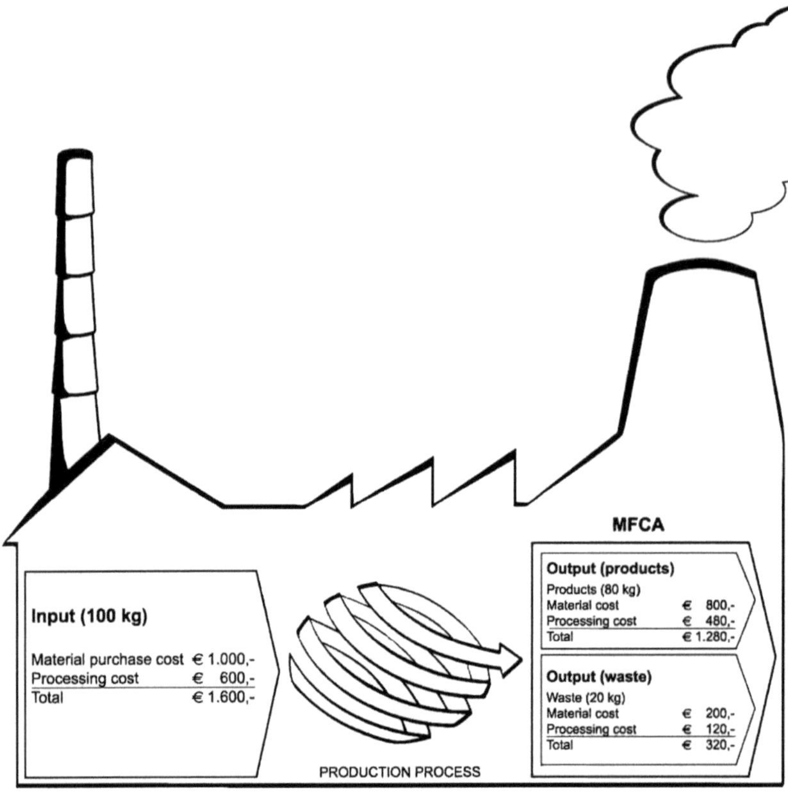

Fig. 7.6 Cost of waste in MFCA

inputs comprise all materials, water and energy, but for specific project, e.g. related with improving product design, the focus may be on raw materials or other material groups only.

While EMA often stays at the corporate or site level, MFCA requires mapping of material flows along cost centers and production processes as described in Chapter 7.5. It can also be applied to specific product manufacturing steps only (Japanese METI, 2007). In quantifying the material flows along defined system boundaries, the organization creates a consistent database containing quantities, values, and costs. The database relates quantities (in physical units like numbers, kg, m^3, kWh etc.), values (= physical quantity × input price) and costs that refer to the material flows (e.g. material costs, inventory values, and waste volumes). In addition, all the other costs incurred by the organization to maintaining the material flow system (e.g. personnel costs, depreciation) are attributed to the respective material flows.

The in-house material flows thus become the core focus of cost analysis and efficiency improvements. MFCA may, for instance, reveal that a measure designed to raise efficiency in a production system results not only in lower material inputs but also in lower costs for materials handling and waste management. Changing to a new colorant, for example, may result not only in different absorption levels, but also in reduced costs for water treatment.

MFCA thus distinguishes the following cost categories (Strobel, 2000; LfU, 1999):

- Material Costs
- System Costs and
- Delivery or Disposal Costs

Material costs

MFCA, as well as the IFAC and UN DSD EMA approach, start with the collection of the physical quantities of materials involved in the various production processes and inventories. The volumes are connected with material purchase prices. The output is distinguished into positive and negative product output. The procedure has been described in previous chapters.

System costs

The material inputs into the various production processes are regarded as cost drivers that have to bear the additional operational costs. "System costs" are by definition all costs that are incurred during in-house handling of materials (e.g. personnel costs, depreciation). System costs allocated to material flows are defined as "system values". Whether these flows are raw materials, intermediate or semi-finished goods, or material losses, each in-house material flow can be seen as a cost carrier, which has to carry its system costs. System costs are allocated to the outgoing product flows (e.g. from the "production" cost center) and then passed on as system values to the subsequent flows and inventories.

Delivery or disposal costs

All outputs have to carry either delivery or disposal costs. They include all costs incurred in ensuring that positive and negative products leave the company, i.e. not only transport costs for products but also costs for disposing of waste and the fees for waste water and effluent control.

The result of MFCA is a complex cost accounting tool showing quantities, values, and costs of material flows, separated into the three categories "material", "system", and "delivery or disposal". Experiences from projects primarily in Germany (Strobel, 2000; LfU, 1999; Strobel and Redmann, 2001; Wagner and Enzler, 2006) and Japan (Kokubu and Nakajima, 2004; Kokubu/Nashioka, 2005) show that this can bring about fundamental changes in a company's way of seeing things, of making decisions, and of acting, whereas in traditional cost accounting, after the first processing stage when the intermediate product is calculated, material costs and system costs are already mixed together. It thus very soon becomes impossible to list costs and values separately according to the three categories either for material movements or for inventories.

Experience reported (Strobel, 2000) also show that a company's existing database, material management system and production planning and control system, will usually contain the majority of the data needed. The extra effort and expense involved in implementing flow cost accounting is thus not so much the continuous tracing of additional data but rather the system's one-time installation.

Benefits of MFCA are

- Cost-reduction and environmental impact reduction as a result of improved material and energy efficiency (i.e. reduced residual waste and reduced use of materials per product)
- Incentives to develop new products, technologies, and procedures
- Enhanced quality and consistency of corporate information systems, linking physical and monetary data
- Improvement of organizational structures and procedures as a result of consistent referencing to the material flow system
- Inter-departmental, material-flow-related communication and coordination instead of separation into divisions, departments, and cost centers with separate responsibilities
- Increased motivation in staff and management regarding the comprehensive structuring of material flows and
- Focus on raising material and energy productivity instead of reducing the workforce

7.8 Investment Appraisal and Budgeting

Managers face a typical dilemma when it comes to investment decisions related with environmental protection. On the one hand, regulatory requirements, voluntary standards, price developments for energy and materials as well as market pressures impose continually higher, and more costly, demands for environmental performance and material efficiency. On the other hand, the information needed for a cost-effective response to such demands is typically unavailable in a timely, rigorous, and consistent way (Jasch 2007). The result is that decisions on investment and management projects, materials choices, product pricing, and product mix often serve neither the best interests of the firm nor the environment.

Most companies have problems quantifying the cost savings of environmental management systems and other environmental activities. Companies generally calculate the cost savings of environmental management by comparing waste streams before investments and good housekeeping measures with later disposal and other costs, or by comparing old and new performance indicators and calculating the difference in monetary value (see the example of the brewery Murau in Chapter 8). Most of these calculations are based on the question: What would I have to pay today if I hadn't invested or acted a couple of years ago? Future oriented calculations additionally face information gaps and uncertainty to answer the question: What will I have to pay in the future if I don't invest or act now?

Conclusions from several case studies emphasize the need for

- Improved consistency between physical and monetary data and related departments
- Material flow accounting as a basis for good cost accounting and
- Adequate consideration of less tangible costs for the calculation of investment appraisal

7.8.1 Capital Budgeting Basics

The basic idea of capital budgeting is to compare different investment alternatives. Investment appraisal is used to determine the cost savings of an investment with regard to its goals. The economical variables for assessment in static financial analysis include

- Initial investment costs
- Operating costs and earnings
- Profit
- Return on Investment and
- Pay-back period

All methods of investment appraisal assume that all future inputs and outputs of an investment decision are quantifiable and can be monetarised.

In dynamic financial analysis, the expected future monetary inflows and outflows are discounted to the time of the investment and calculated into internal discount rate or annuity. The opportunity costs of capital (the lower value of cash flows which don't occur today, but only in the future) are considered by discounting them with the interest rate of financial markets. The sum of all discounted future cash flows determines the net present value of a project or investment, which is compared to the value of the old equipment and to the interest rate of financial markets. A planned investment has to be more profitable then gaining interest on a bank deposit.

Payback methods for capital budgeting do not consider cash flows beyond the payback period. Many companies adopt internal rules, that only projects with a payback period of two or three will be accepted, regardless of possible longer term benefits. Discounted cash flow methods in principle consider all relevant future cash flows until the project ends, but as many companies apply excessively high interest rates, which result in a negligible present value for medium and long term costs and savings, only the first 3 years count in effect for the investment decision.

The approach and shortcomings of methods such as the payback period, internal rate of return, or internal interest rate (IIR) are discussed in any textbook on corporate finance.

The methods for determining the value of a company for mergers and acquisition are also based on capitalized future earnings. Low environmental risks and the capacity to respond to future trends and stakeholder demands can increase the value of the company.

The high risks, difficult monetarization and high uncertainty of many environment related future costs, as well as the potential cost savings of cleaner technologies arising from the reduced use of hazardous auxiliary and operating materials and related environmental protection measures have made estimation of the future even more difficult. Still, the methods are widely used. The task is not so much to change the basic concept of discounting future monetary flows, but to ensure the inclusion of all relevant earnings and expenses.

Quantifying future earnings and output flows resulting from measures for environmentally protection is a difficult venture. Particularly in the area of environmental management, it is necessary to include "soft" or less tangible values. In addition to pure investment and operation costs, factors such as image, future liability costs, future price changes for raw materials and energy carriers and their availability, contacts with environmental and other agencies, legal compliance, employee motivation etc. need consideration.

Initial investment costs can comprise several cost categories in addition to the purchased equipment. Annual operating costs can relate to all the other cost categories of the environmental cost scheme. Therefore, annual assessment of total expenditure is vital as a starting point in environmental management accounting. It assures the once complete picture, which later allows consideration of the details required, related to specific cost centers or equipment.

Measures for pollution prevention help to reduce not only disposal and emission treatment costs but often also increase the efficient use of purchased materials and energy. Often, when calculating investments, the reduced costs for materials and emission treatment are not completely calculated. This results in distorted investment decisions. In addition, future liability costs and less tangible benefits should be estimated. They may comprise liabilities for personal injury or property damage (e.g. liability stemming from a leaking landfill), and penalties and fines for violation of environmental regulations. To the degree that a clean up obligation is legally required, a provision has to be made in the balance sheet.

The following saving potentials may be considered:

- Cost reduction for waste and emission treatment and disposal costs. This includes internal and external treatment, related equipment and operating materials, personnel handling waste, storage and landfill costs, fees for disposal, transport, insurance and liability, site and production permits, reports to authorities, etc.
- Savings of insurance, liability and remediation costs. Reduced waste and emissions and new processes requiring less harmful operating materials are often

also a good argument for reduced risks of damage, spills, land contamination, cleanup-obligations or other possible remediation costs.
- Maintenance: Labor and material for maintenance can also be affected by product design and cleaner technologies.
- Savings in energy and water input: Generally, cleaner technologies not only require less material input but are also more energy and water efficient.
- Savings in raw and auxiliary materials and packaging. Alternatives which reduce the amount of waste, in general also need less material input.
- Savings because of better product quality. Alternative product design can improve the product quality and thereby reduce the costs of quality control, redoing work and production of scrap.
- Earnings from new by-products. If waste is replaced by new, marketable by-products, the cost of new product design can be offset by those earnings.
- Reduced risk of accidents and worker absenteeism by avoiding dangerous materials and processes which also results in increased employee motivation.
- Improved relations with local authorities speed up the time required for production permits and other official procedures.
- Future investment savings through anticipation of planned policy changes (i.e. stricter emission allowances, prohibited use of hazardous materials), thus preventing the requirement for short term or end-of-pipe solutions.

In addition to savings, environmental management systems have resulted in a lot of other positive effects, such as

- Increased turnover, customer satisfaction, new markets, differentiation from competitors
- Better Image and product branding
- Better relations with authorities, reduced regulatory compliance costs
- Better creditworthiness with banks, reduced insurance rates, good ratings by investment brokers and agencies
- Better public stakeholder and community relations
- Increased job motivation and satisfaction, less absenteeism and worker illness

7.8.2 Calculation Sheet for Environmental Investments and Savings

The calculation sheet for investments and projects in Table 7.7 can be used to calculate several alternatives and comparing them, or to directly estimate resulting cost savings. An annual assessment of total environmental expenditures should have been performed beforehand, in order to provide a sound data basis. Depending on the project or investment, only some columns and rows may be filled, but the likelihood of forgetting significant cost factors is decreased.

Table 7.7 Calculation sheet for environmental investments and projects

Calculation sheet
Applied separately for the existing and planned equipment or directly for calculated savings

Environmental cost categories	Initial investment	Year 1	Year 2	Year 3	Year 4ff
Material purchase value of non-product output					
Raw materials					
Packaging					
Auxiliary materials					
Operating materials					
Water					
Energy					
Processing costs of non-product output					
Waste and emission control costs					
Depreciation for related equipment					
Operating materials					
Water and energy					
Internal personnel					
External services					
Fees, taxes and permits					
Fines and penalties					
Insurance					
Remediation and compensation					
Prevention and other environmental management costs					
Depreciation for related equipment					
Operating materials, water and energy					
Internal personnel					
External services					
Other environmental management costs					
Research and development					
Σ Environmental costs					
5. Environmental earnings					
5.1. Subsidies, investment grants					
5.2. Other earnings					
Σ Environmental earnings					

Allocation to the different environmental media will probably not be necessary, so the columns have been modified to time series.

Once the total environmental costs of two alternatives have been assessed for 1 year, they can be extended into time series for capital budgeting. Estimates of monetary inputs and outputs for the first 3 years should be more detailed. For years 4 to 10 rough annual estimates would be sufficient.

The determination of total annual environmental expenditure for the last business year is a prerequisite for calculating options. If the total annual environmental costs have not been assessed, the savings potential can't be calculated. After the determination of the total annual environmental costs, the calculation can be done for specific cost centers or production processes. Calculating different options is then relatively straightforward.

When comparing investment options, it is advisable to first assess the cost of the old equipment with the proposed scheme and then calculate the costs of the new equipment. The so-called soft factors or less tangibles, can be included in the investment decision process, if necessary.

7.9 Benchmarking Production Sites

In 2005 and 2006, Danisco A/S carried out a corporate pilot program, "Global Waste Initiative", for testing the adequateness of IFAC's guidance document on environmental management accounting (EMA) as a tool for production sites in the global biotech and food ingredients industrial sector (Munkoe and Jasch, 2008). The pilot sites, that participated in the case study, were diverse from a geographical and production process point-of-view in order to demonstrate differences and similarities.

The objectives of the pilot project were

1. To investigate EMA as a strategic cost assessment tool for subsequent identification and evaluation of environmental saving initiatives
2. Comparison of EMA results versus annual reported environmental costs for production sites
3. To evaluate EMA as a benchmarking tool and
4. To evaluate required resources for future EMA assessments

The main conclusions of the three pilot assessments were that the annual environment-related costs are considerably higher than the recordings of the individual sites before the EMA project. In addition the assessments emphasized the need for strengthening the relation between the environmental and accounting information systems of the manufacturing sites in order to get a complete picture for decisions regarding improvement of energy and material efficiency.

Danisco is a global business-to-business supplier of enzymes and food ingredients based in Copenhagen, supplying customers from more than 70 manufacturing facilities throughout the world and more than 10,000 employees world-wide. A global program for waste and wastewater reducing initiatives was launched in 2005. The related EMA pilot assessments were conducted at three of the manufacturing facilities in Finland, France and USA, respectively. The intention was to illustrate EMA as a strategic cost assessment tool for identification of environmental saving initiatives.

The assessments revealed that several environmental costs were unknown to operational management, as they had been posted to accounts where nobody expected them and looked for them. Thus, when trying to come up with the total environmental costs, e.g. for calculating investment options or external disclosure purposes, they got lost.

Direct costs posted to a production cost center or product, could be traced comparatively easy, once the related cost center or product had been identified as "environmentally relevant". But all indirect costs posted to general overhead accounts were practically impossible to be traced later onwards, as the accounts didn't contain remarks on the separate bookings, but simply invoice numbers within further text.

The environmental costs revealed by the EMA methodology may have been posted to several cost centers and accounts, but often got lost during aggregation as the information flow between the different departments and information systems is not non-functioning. For the EMA assessments, a team consisting of the environmental manager, a production manager and people working with cost accounting/controlling was put together. Each assessment required approximately three internal persons per site for two days to complete an EMA assessment including a first material flow balance for the previous business years. Nevertheless, all teams stated that future assessments would be uncomplicated as the workflow and information sources have been identified. The three sites estimate that approximately one half day only will be needed for future assessments. In addition, the consistency and comparability of data was improved significantly. Now, the environmental manager knows what to look for, and the accountants will help where to find data.

The environmental costs were analyzed by cost categories and environmental domains and the production sites benchmarked against the average cost distribution, which is shown in Table 7.8. As EMA includes all the energy purchase as environmentally relevant, this cost category normally constitutes a significant share of environmental costs. For the DANISCO sites the total energy purchase and resulting impact on air and climate accounts for 52% of the total costs by environmental domain. The other important environmental domains impacted are water/waste water as well as solid waste with 24% and 23% of total costs. General environmental management accounts for 1% only. As often, several of the columns requested by national statistical reporting of environmental costs are not relevant for this business sector (soil and groundwater protection, noise, dust, vibration, biodiversity and radiation).

The analysis by cost categories revealed the importance of the NPO approach. Materials costs of non-product output account for about 88% of the total average annual environment-related costs of the three assessments. This highlights the fact that when comparing the costs of non-product materials with the costs of environmental protection and management, the latter is comparatively negligible. Waste

Table 7.8 Average distribution of DANISCO's environmental costs

Environmental domain **Average percentage distribution of 3 Danisco sites** **Environmental cost categories**	**Air + climate**	**Waste water**	**Waste**	**General environmental management**	**Total**
Materials costs of products					
Raw and auxiliary materials				87	87
Packaging materials				5	5
Operating materials				2	2
Energy				6	6
Total materials cost of products				**100**	**100**
Material costs of non-product output	**52**	**15**	**22**		**89**
Raw and auxiliary materials	2	7	5		14
Packaging materials			1		1
Operating materials	2	5	14		21
Water		3			3
Energy	48				48
Production costs of NPO			2		2
Waste and Emission Control Costs	**1**	**8**	**1**	**1**	**11**
Depreciation		1			2
Water and energy		5			5
Personnel		1			1
Taxes, fees, permits		1	1		2
Prevention and environmental management cost				**1**	**1**
Personnel				1	1
Σ Total environmental costs	**52**	**24**	**24**	**1**	**101**
5. Environmental earnings			**–1**		
Σ Total environmental costs and earnings	**52**	**24**	**23**	**1**	**100**

and emission control costs account for 11% of total costs, while prevention makes up only 1%, while it helps save costs in the other cost categories. The costs for emission control are mostly connected with waste water treatment and related equipment, water input, energy costs and personnel that could sometimes be taken directly from the cost center for the waste water treatment plant. Disposal fees, waste water treatment fees and related permits account for only 2% of total costs.

Prevention costs in the assessments mostly consist of costs for internal personnel in the environmental management department plus external consultants dealing with specific projects.

Before conducting the EMA assessments on sites all sites were requested to report their environmental costs and energy consumption as a starting point. All sites already annually collected the costs for waste, wastewater and energy. But the EMA assessment clearly defines environment-related costs in the categories for control and prevention from different perspectives and takes into account the costs of losses. For this reason the EMA environment-related costs differs considerable from the usual way of making up the environmental costs for the sites. The EMA costs assessed at the three sites were 170%, 180% and 245% above the costs reported at project start.

Both the local sites management teams and the divisional management at Danisco found EMA excellent as a benchmarking tool.

Compared to current practice, the value added by EMA for the site management is cost monitoring over time by environmental domains. The detailed picture of each element of the environmental costs enables both the management and the production organization to improve performance and thus reduce the environmental impact. The inclusion of the costs of material losses (NPO) was found especially useful for benchmarking environmental performance and measuring continuous improvements. All sites assessed had environmental management systems installed which ensure a focus and commitment of the management to improve the environmental impact from the site. EMA in addition provides the link between the environmental management system and the financial information system. Benchmarking between the sites, and benchmarking the individual sites over time, will reveal differences of operation and technology platforms and such inspire for improvements. One of the sites had considerable higher costs for energy and consequently, an energy audit was performed, which resulted in considerable energy savings (Munkoe and Jasch, 2008).

The project results confirmed EMA as a suitable a benchmarking tool, also between production sites and for identifying cost flows in production over time. As future assessments of the individual sites will reveal the development in environment-related costs, new focus areas will be discovered. Benchmarking between sites using a comparable technology platform was also interesting from a management perspective in spite of cultural and regional differences.

The subsequent evaluation with the local site management revealed important aspects regarding the interfaces between the accounting and operational site management, and the information systems and procedures related to production and environmental control. In general, production has a considerable focus on the reduction of material losses and product yield in all cases. But, in spite of this, the related costs of material losses identified by EMA were not visible for production in the daily work.

Similarly, the focus of the environmental management department was mainly on environmental control and impacts and less on related costs.

An important result of the project was the confirmation that EMA offers a strengthened linkage between the environmental management system and operational management by providing integration of information systems from management, production, accounting and environment.

The total environment-related costs in each assessment far exceeded the perception of the local organizations. By presenting an alternative and detailed cost structure for the environmental domains and usually increased environment-related costs, operational management was offered a more precise tool for evaluating investments and environmental initiatives. As a consequence, management may improve both environmental and financial performance when prioritizing environmental focuses and setting environmental targets.

Chapter 8
Case Study of a Brewery

Chapter 8 describes a case study developed from the brewery Murau in depth and at the same time demonstrates how to use the excel template for the EMA cost assessment that is provided as a download under www.ioew.at and www.springer.com/978-1-4020-9027-1

8.1 Working with the EMA Excel Templates

This chapter contains a description on how to work with the EMA excel templates. A detailed assessment aid in an Excel template that follows Tables 1 and 2, (Executive Summary), 2.1 and 6.1 is available for download under www.ioew.at and www.springer.com/978-1-4020-9027-1.

The EMA Excel-template consists of four sheets–*Mass balance, Detail, Sum, and Structure*. Information is only added into the *Mass balance* and the *Detail* sheet.

The *Mass balance* records the physical and monetary values of material inputs and product outputs in one work step, as these amounts should be consistent. The excel template contains two columns for the source of information for both values. The enterprise resource planning system and the accounts for materials used for production should provide this information in a consistent and detailed manner.

For product output only the volumes, but no monetary values are collected, as these costs are assessed later in the cost category waste and emission treatment fees.

The mass balance is not automatically calculated, as in most organizations the data necessary is not available for the first assessment and depending on the production process adjustments may be needed. Companies may find it useful to separately calculate the mass, the energy and the water balance with the help of their process technician. The actual cost assessment is performed in the *Detail* sheet only.

All the cost categories are already set but the several different cost items related to cost accounts or taken from cost center reports should be listed with indicating the reference. The environmental domains can be modified if necessary. If columns are added or deleted, then the same has to be done for the other two sheets.

C. Jasch, *Environmental and Material Flow Cost Accounting*: Principles and Procedures, Eco-Efficiency in Industry and Science 25,

The program automatically aggregates the costs of each cost category, but when adding lines to fill in more details a last cross check is recommended to make sure all aggregates are complete.

The sum of the costs of all categories in the sheet *Detail* is automatically transferred to the sheet *Sum* to have an overview and a better presentation layout. The sheet *Structure* merely calculates the costs into percentages to show the most relevant environmental costs.

It is recommended for costs that are incurred by defined equipment to simultaneously collect the data on maintenance, external services, personnel, and material costs, especially if this information is available from cost center reports. Care needs to be taken to avoid double counting, if e.g. operating materials are collected from cost center reports under the cost category for emission control or integrated prevention and from accounts under the cost category for NPO or if external services are taken from expenditure accounts and costs centers as well.

All collected data should be assigned to the correct environmental domain (media) or to general environmental management. Some companies have also added columns, e.g. for health and safety or for product oriented prevention activities.

The column *Account* is to keep record of the cost centers and accounts for the years to come without having to spend a lot of time finding them again. It is also practical to document the type of calculation used to acquire a certain figure. It is possible to add lines into the sheet, just beware of maintaining the automatic excel calculations.

The sheet includes a control function, which ensures that the value in column *Costs in €* is identical to that of *Sum*. If not, an error is displayed. The values are only identical if all costs in the *Costs in* are assigned to a domain.

8.2 The Material Flow Balance

The case study for a fictitious brewery has been developed based on a real pilot project with the brewery in Murau in Austria. It shows the result of a one day assessment at a typical brewery with about 150 employees and a good environmental management and indicator system. It specifically tries to show how the excel templates are being filled out. The values have been slightly modified.

Table 8.4 also allows a better understanding of the different postings in the detailed environmental cost assessment. Starting point of an EMA assessment is the recording of materials flows (Chapter 2 & 3, Tables 2.1 and 8.1). It is followed by the detailed EMA cost assessment, applying the EMA Excel template sheet *Detail* (Tables 8.2 and 8.5). In order to make use of the total annual environmental cost for investment appraisal, it might be useful to record the environmental costs by cost centers or production processes (Chapter 7).

Tables 8.3 and 8.4 show how the detailed EMA cost assessment is automatically aggregated into a one page spreadsheet and a percentage distribution of costs. These Excel templates are most useful of interpretation of results and monitoring of changes during subsequent years.

Table 8.1 Input output framework of the brewery

Input	Ouput
Raw materials	**Product**
Barley	Bottled beer
Wheat	Beer in kegs
Malt	Canned beer
Hops	Alcohol-free drinks
Brewing water	**By-products**
Auxiliary materials	Malt
Additives (beer)	Malt dust
Additives (lemonade)	Hops
Laboratory materials	Barley waste
Packaging	Spent grains
Crates (new)	Silicic acid
Bottles	**Waste**
Cans	*Total waste for recycling*
Kegs	Glass
Palettes	Metal
Labels	Etiquettes
Foil	Plastics
Corks	Paper, cardboard
Caps	*Total municipal waste*
Operational materials	*Total hazardous waste*
Cleaning materials	Fluorescent tubes
Disinfecting materials	Refrigerators
Neutralisers	Oils
Filters	Oil contaminated materials
Oils/grease	Used inks
Salts	Chemical remnants
Cooling materials	Electrical scrap
Repair and maintenance materials	**Waste water**
Canteen	Amount in m^3
Office	COD
Other	BOD
Energy	Phosphates
Gas	Nitrogen
Coal	Ammonium
Fuel oil	
District heat	**Air-emissions**
Renewables (Biomass, Wood)	CO
Solar, Wind, Water	CO_2
Externally purchased electricity	SO_2
Internally produced electricity	NO_x
Water	
Municipal Water	**Noise**
Ground water	Maximum Noise at night
Spring water	Maximum Noise on site
Rain/ Surface Water	

Table 8.2 Process flow chart of the brewery

Input	Main process	Side processes	Output
Malt Energy	Grinding		Dust
Brewing water Detergent Energy	Mashing		Heat
Water Energy	Purification		Spent grains Heat Waste water
Hops Energy	Preparation of wort		Heat
Water Energy	Removal of hops waste		Hops waste
Water Energy Detergent Refrigerant	Cooling of wort		Warm water
Yeast Sterile air Water Energy Refrigerant	Fermentation		Yeast Wasted beer Carbonic acid Waste water
Water Energy Refrigerant Disinfectant	Storage		Storage dust Waste water Wasted beer CO_2
Water Energy Carbonic acid Detergent Disinfectant Auxiliary materials	Filtration		Waste water Filtrate Auxiliary materials
Water Energy Refrigerant Detergent Disinfectant Carbonic acid	Pressurization		Waste water CO_2
Water Energy Detergent Disinfectant Bottling		Bottle and Cask cleaning	Waste water Waste paper Waste glass Sludge Heat
Lemonade raw materials Sugar		Lemonade production	

(continued)

Table 8.2 (continued)

Input	Main process	Side processes	Output
Water	Bottling, Casking		Bottled wasted beer
Energy			Casks, boxes
Carbonic acid			Packaging waste
Packaging			Waste glass
			Rinsing water
			Residue
			Waste water
Department specific Inputs		Workshop, canteen, administration	Department specific Outputs
Fuel oil		Steam/heat production	Air emissions
Water			
Petrol	Transport and delivery		Air emissions

8.3 The Brewery, Its Production Flow and Cost Centers

Obermurtaler Breweries is a small country side brewery with about 150 people. It has implemented ISO 14001 and EMAS (European Comission 2001b) for 12 years and was actually the first Austrian site to be EMAS verified. It carries the Austrian Ecolabel for returnable beer bottles. It has also participated in pilot studies to develop the UNDSD and IFAC EMA approach. The following data is based on the extensive environmental statement for 2005 (www.murauerbier.at) and pilot studies, where also other breweries were involved (Jasch and Schnitzer, 2002). The data does not directly relate to the actual figures of the brewery.

The total annual environmental costs are assessed together with an extensive performance indicator system on an annual and partly monthly basis. The environmental costs are traced from the list of accounts, the cost center reports and performance indicator reports from production statistics (e.g. materials input per beer produced, loss percentages and production volume) and environmental management (e.g. waste volumes).

Tables 8.1 and 8.2 show the Material Flow Balance of the brewery. The physical mass balance doesn't balance off to zero, as not all volumes are recorded yet (e.g. packaging volumes, tools and maintenance supply). As water is part of the product, the mass balance is rather tricky, having to include the energy and water balance as well. Care should also be taken not to aggregate different measurement units (tons, m^2, m^3, pieces, etc.) But even without balancing the input output analysis provides a very good controlling instrument and figures are monitored for each relevant material group on a separate account.

The monetary value of non-product output is traced in the subsequent assessment of financial data, but not in the mass balance. Turnover needs not be accounted for EMA purposes.

The focus in recent years has been to record also operating materials in the enterprise resource planning system and record their consumption volumes also on a cost center level in order to be able to better monitor material flows.

8.4 Total Annual Environmental Costs

The template in Table 8.5 shows a detailed practical example of how environmental costs are recorded in the Excel template following the UN DSD and IFAC EMA Guideline. For the EMA cost assessment all postings are entered only into this spreadsheet which automatically aggregates to the results presented in Tables 8.3 and 8.4.

The assessment normally doesn't take more than half a day to a day, working in a team consisting of the environmental manager, the accountant with direct access to the cost accounting system and the production manager. It is essential to record the source of information and the procedures for estimates in order to be able to repeat the cost assessment in a comparable way with less effort next year.

For the EMA assessment much of the data will be taken directly from cost center reports of defined environmentally relevant end-of-pipe or pollution prevention equipment. It may be useful to monitor these cost centers in separate columns in addition to the distribution by environmental media affected.

For investment appraisal it may be sufficient to first record the total annual environmental and material flow related costs and than separate only those costs centers, for which investment appraisal will be performed.

The detailed cost assessment is automatically aggregated into a one page display of the totals of the sub-cost categories (Table 8.3). In many companies the columns requested for reporting to statistical agencies for

- Soil, surface and ground water
- Noise, vibration, odor and fire, as well as
- Nature protection

Remain empty. The interpretation of results is simplified by referring to the automatically converted excel template of the percentage distribution of the total annual environmental costs (Table 8.4).

The percentage distribution of total annual environmental costs clearly shows that emission control costs are comparatively expensive in relation to prevention activities. But even in a company that has practiced environmental management and integrated prevention for 20 years, the most significant cost category are the materials costs of non product output with 67% of total costs. This is where one still finds saving potentials.

On the other it must be said, that price changes also influence these figures. In the light of rising resource prices many companies are horrified by the thought of what they would have to pay today had they not invested into efficiency improvements in the last years. It must also be said that for the brewery for 2002 total energy input already constituted 27% of environmental total costs.

Several companies don't publish their actual cost but do disclose the percentage distribution. The figure for energy provides a good estimate of the total relation of the cost structure. Energy related impact on air and climate is also the most important cost category by environmental media.

The next two significant cost items are the losses of raw materials and operating materials. Together they are in the range of total energy input. While raw materials are more commonly monitored by organizations, the recording of operating materials by production processes and cost centers is not so common.

Only 2.7% of the total costs related to the operating materials directly attributed to the waste water treatment pant (line 2.2) but another 11% of total costs related to operating materials that go down the drain (cleaning materials, lubricants, detergents, etc.).

When analyzing the cost distribution by environmental domains it is interesting to note that for the brewery in Austria in recent years the most prominent category shifted from waste to waste water and now stands at air and climate. This clearly relates to priorities of environmental politics and related price changes. Much of the solid waste is recycled and some is even sold which shows in line 7.1, Other earnings.

Table 8.3 Total annual environmental costs of the brewery

Environmental domains Environment related cost categories	Air and climate	Water + waste water	Waste	Soil, surface and ground water	Noise, vibration and odour and fire	Nature protection	General environm. MS	Total
1. MATERIALS COSTS OF NON-PRODUCT OUTPUTS	**782,100**	**385,243**	**514,214**	**0**	**0**	**0**	**0**	**1,681,557**
1.1. Raw and auxiliary materials	100,000	10,243	228,030					338,273
1.2. Packaging materials			240,184					240,184
1.3. Merchandise	0	0	0	0	0	0	0	0
1.4. Operating materials		275,000	46,000					321,000
1.5. Water		50,000						50,000
1.6. Energy	682,100							682,100
1.7. Processing costs		50,000						50,000
2. END-OF-PIPE	**40,540**	**262,800**	**229,000**	**5,730**	**0**	**0**	**60,500**	**598,570**
2.1. Equipment depreciation	1,240	89,100	30,000	1,030				121,370
2.2. Operating materials	7,000	68,200						75,200
2.3. Water and energy	5,000							5,000
2.4. Internal personnel	27,300		125,000				40,000	192,300
2.5. External services		500	10,000	3,000			20,500	34,000
2.6. Fees, taxes and permits		105,000	64,000					169,000
2.7. Insurance								
2.8. Remediation and compensation				1,700				1,700

3. INTEGRATED PREVENTION	**37,300**	**500**	**0**	**0**	**0**	**0**	**237,800**	**275,600**
3.1. Equipment depreciation	37,300	500						37,800
3.2. Operating materials, water, energy								
3.3. Internal personnel							222,500	222,500
3.4. External services							10,300	10,300
3.5. Other							5,000	5,000
4. RESEARCH and DEVELOPMENT COSTS	**10,000**	**0**	**0**	**0**	**0**	**0**	**0**	**10,000**
5. FINES	**0**	**0**	**0**	**0**	**0**	**0**	**0**	**0**
TOTAL ENVIRONMENT-RELATED COSTS (1. + 2. + 3. + 4. + 5. + 6.)	**869,940**	**648,543**	**743,214**	**5,730**	**0**	**0**	**98,300**	**2,565,727**
6. ENVIRONMENT-RELATED EARNINGS								
61. Other earnings			–38,500					–38,500
6.2. Subsidies	–3,000							–8,000
TOTAL ENVIRONMENT-RELATED EARNINGS	**–3,000**	**0**	**–38,500**	**0**	**0**	**0**	**0**	**–46,500**
TOTAL ENVIRONMENT-RELATED COSTS & EARNINGS	**866,940**	**648,543**	**704,714**	**5,730**	**0**	**0**	**298,300**	**2,519,227**

Table 8.4 Percentage distribution of total environmental costs of the brewery

Environmental domain / Environment-related cost categories	Air and climate (%)	Water + waste water (%)	Waste (%)	Soil, surface and groundwater (%)	Noise, vibration and odour and fire (%)	Nature protection (%)	General environm. MS (%)	Total (%)
1. MATERIALS COSTS OF NON-PRODUCT OUTPUTS	**31.0**	**15.3**	**20.4**	**0.0**	**0.0**	**0.0**	**0.0**	**66.7**
1.1. Raw and auxiliary materials	4.0	0.4	9.1	0.0	0.0	0.0	0.0	13.4
1.2. Packaging materials	0.0	0.0	9.5	0.0	0.0	0.0	0.0	9.5
1.3. Merchandise	0	0	0	0	0	0	0	0
1.4. Operating materials	0.0	10.9	1.8	0.0	0.0	0.0	0.0	12.7
1.5. Water	0.0	2.0	0.0	0.0	0.0	0.0	0.0	2.0
1.6. Energy	27.1	0.0	0.0	0.0	0.0	0.0	0.0	27.1
1.7. Processing costs	0.0	2.0	0.0	0.0	0.0	0.0	0.0	2.0
2. END-OF-PIPE	**1.6**	**10.4**	**9.1**	**0.2**	**0.0**	**0.0**	**2.4**	**23.8**
2.1. Equipment depreciation	0.0	3.5	1.2	0.0	0.0	0.0	0.0	4.8
2.2. Operating materials	0.3	2.7	0.0	0.0	0.0	0.0	0.0	3.0
2.3. Water and energy	0.2	0.0	0.0	0.0	0.0	0.0	0.0	0.2
2.4. Internal personnel	1.1	0.0	5.0	0.0	0.0	0.0	1.6	7.6
2.5. External services	0.0	0.0	0.4	0.1	0.0	0.0	0.8	1.3
2.6. Fees, taxes and permits	0.0	4.2	2.5	0.0	0.0	0.0	0.0	6.7
2.7. Insurance	0.0	0.0	0.0	0.0	0.0	0.0	0.0	0.0
2.8. Remediation and compensation	0.0	0.0	0.0	0.1	0.0	0.0	0.0	0.1

3. INTEGRATED PREVENTION	**1.5**	**0.0**	**0.0**	**0.0**	**0.0**	**0.0**	**9.4**	**10.9**
3.1. Equipment depreciation	1.5	0.0	0.0	0.0	0.0	0.0	0.0	1.5
3.2. Operating materials, water, energy	0.0	0.0	0.0	0.0	0.0	0.0	0.0	0.0
3.3. Internal personnel	0.0	0.0	0.0	0.0	0.0	0.0	8.8	8.8
3.4. External services	0.0	0.0	0.0	0.0	0.0	0.0	0.4	0.4
3.5. Other	0.0	0.0	0.0	0.0	0.0	0.0	0.2	0.2
4. RESEARCH and DEVELOPMENT COSTS	**0.4**	**0.0**	**0.0**	**0.0**	**0.0**	**0.0**	**0.0**	**0.4**
5. FINES	**0.0**	**0.0**	**0.0**	**0.0**	**0.0**	**0.0**	**0.0**	0.0
TOTAL ENVIRONMENT-RELATED COSTS (1. + 2. + 3. + 4. + 5. + 6.)	**34.5**	**25.7**	**29.5**	**0.2**	**0.0**	**0.0**	**11.8**	101.8
6. ENVIRONMENT-RELATED EARNINGS								
6.1. Other earnings	0.0	0.0	–1.5	0.0	0.0	0.0	0.0	–1.5
6.2. Subsidies	–0.1	0.0	0.0	0.0	0.0	0.0	0.0	–0.3
TOTAL ENVIRONMENT-RELATED EARNINGS	**–0.1**	**0.0**	**–1.5**	**0.0**	**0.0**	**0.0**	**0.0**	**–1.8**
TOTAL ENVIRONMENT-RELATED COSTS & EARNINGS	**34.4**	**25.7**	**28.0**	**0.2**	**0.0**	**0.0**	**11.8**	**100.0**

Table 8.5 Detailed EMA cost assessment in the Excel template

Process Flow Chart

Input	Production CC	Supportive CC	Output
		Storage facilities for brewing and operating materials including CIP plants	
Malt, brewing water, cleaning agents, energy	Brewing malt and mills (grinding, mashing and purification)		Spent grains, dust, heat, waste water
Hop, water, cleaning agents, detergents, energy, refrigerant	Brew house, wort production (Stammwürze)		Hops waste, brewing residue, heat, waste water
Yeast, sterile air, refrigerant, water, energy	Fermentation and storage cellar (fermentation of the malt sugar with yeast)		Yeast, wasted beer, carbon dioxide, waste water
Water, energy, carbonic acid, cleaning agents, disinfectants, refrigerant, auxiliary materials	Filtration (separation of yeast and proteins)		Waste water, filtrate, auxiliary materials, carbon dioxide
Water, energy, carbonic acid, cleaning agents, disinfectants, packaging materials	Bottling and barrel filling		Waste water, sludge, solid waste, heat, residue, bottled wasted beer
Operating materials, energy		Maintenance	Operating materials
Energy, refrigerant		Steam/heat production	Heat, air emissions
Refrigerants, energy		Refrigeration	Air emissions
Operating materials, energy		Waste water treatment	Waste water, waste
Petrol		Logistics	Air emissions
Operating materials, energy		HSEQ MS	Operating materials
Operating materials, energy		Administration	Operating materials
Total cost centres	5	8	

I-O Balance

MATERIAL Flow Balance/ INPUT/OUTPUT	EUR	tons (unless otherwise indicated)	Source of information for EUR	Source of information for tons
1. Materials inputs			Account number	
1.1. Raw and Auxiliary Materials				Enterprise resource planning system
Malt	1,000,000	4,000	5100	
Hop	120,000	500	5101	
Burst rice	120,200	200	5102	
Auxiliary materials	12,150	100	5110	
CO2 Purchase	100,000		5111	
Subtotal	**1,352,350**	**4,800**		
1.2. Packaging materials				Not yet recorded in volumes
Bottle caps lemonades	17,000		5301	
Bottle caps beer	80,000		5302	
Labels beer	100,000		5310	
Label glue	15,000		5330	
6 bottle-trays	160,000		5340	
Beer bottles	45,000		5341	
Pallets	14,200		5350	
Subtotal	**461,200**	**0**		

(continued)

Table 8.5 (continued)

I-O Balance				
MATERIAL Flow Balance/ INPUT/OUTPUT	EUR	tons (unless otherwise indicated)	Source of information for EUR	Source of information for tons
1.3. Merchandise not to be recorded				
Subtotal	**0**	**0**		
1.4. Operating materials				
Cleaning agents	190,000	210	5400	Enterprise resource planning system
Refrigerants	40,000	50	5401	
Neutralisation agent	35,000	250	5402	
Filtering agents	20,000	30	5403	
Laboratory material	20,000	1	5404	
Lubricants	11,000	1	5405	
Tools and maintenance supply	5,000		5500	Not yet recorded
Subtotal	**321,000**	**542**		
1.5. Water				
Ground water consumption in hl	0	0		Not in use
Water from own wells in HL	0	1,300,000		Metering system
Water consumption from public supply (hl)	50,000	1,000,000	5650	Invoice
Subtotal	**50,000**	**2,300,000**		

1.6. Energy				
Electricity (kWh)	275,000	2,700,000	5600	Invoice
Heating oil extra light (L)	200,000	700	5601	Invoice
Fuels (L)	21,300	300	5602	Invoice
Diesel vehicle fleet (L)	200,000	370,000	5603	Invoice
Subtotal	**696,300**			
TOTAL MATERIALS COSTS/ INPUT	**2,880,850**			
2. Product output			Account number	
2.1. Products				
beer (in hl), bottled or in KEGs	1,000,000	260,000	Total production costs from financial statistics and calculation sheet for production costs	Production statistics
Subtotal	**1,000,000**	**260,000**		
2.2. Byproducts				
Brewing residue for agricultural composting	–3,500	280	4101	Production statistics
Semi-solid Kieselgur mineral silt for agricultural composting	0	240	Delivered free of charge	Production statistics
Wet Draff for agricultural composting	–35,000	5,500	4100	Production statistics
Subtotal	**–38,500**	**6,020**		
TOTAL TURNOVER/ PRODUCT OUTPUT	**961,500**	**266,020**		

(continued)

Table 8.5 (continued)

I-O Balance				
MATERIAL Flow Balance/ INPUT/OUTPUT	EUR	tons (unless otherwise indicated)	Source of information for EUR	Source of information for tons
3. Non-product output				
3.1. Solid waste				
Total non hazardous waste		20		Waste recording system
Waste for recycling		430		Waste recording system
Subtotal		**450**		
3.2. Hazardous waste				
Hazardous waste		7		Waste recording system
Waste oil		0		Waste recording system
Subtotal		7		
3.3. Waste water				
Quantity of waste water in m^3		96,200		Metering system
COD		153		Calculated from laboratory results
Subtotal				
3.4. Air emissions				
CO_2 emissions heating plant		2,500		Calculated from energy input
CO_2 emissions vehicle fleet		1,000		Calculated from energy input
Subtotal		**1,000**		
TOTAL NON-PRODUCT OUTPUT				

Environmental costs detail

Environmental domain		Operating costs (current expenditures)							
Environment-related cost categories	Data source	Air and climate	Water + waste water	Waste	Soil, surface and groundwater	Noise, vibration and odor and fire	Nature protection	General environm. MS	Total
1. MATERIALS COSTS OF NON-PRODUCT OUTPUTS									
1.1. Raw and auxiliary materials									
Malt, 20% loss of €1,000000	5100			200,000					200,000
Hop, 20% loss of €120.000	5101		10,000	10,000					20,000
Burst rice, 15% loss of €120.200	5102			18,030					18,030
Auxiliary materials beer 2% loss of €12.150,	5110		243						243
CO_2 Purchase 100%	5111	100,000							100,000
Subtotal		**100,000**	**10,243**	**228,030**	**0**	**0**	**0**	**0**	**338,273**
1.2. Packaging materials									
Bottle caps lemonades 5% loss	5301			850					850
Bottle caps beer 5% loss	5302			4000					4000

(continued)

Table 8.5 *(continued)*

Environmental costs detail									
Environmental domain		Operating costs (current expenditures)							
Environment-related cost categories	Data source	Air and climate	Water + waste water	Waste	Soil, surface and groundwater	Noise, vibration and odor and fire	Nature protection	General environm. MS	Total
Labels beer 7%	5310			7000					7000
Beer cases 100% of new purchase to the closed loop system	5320			30,000					30,000
Label glue 7%	5330			1,050					1,050
6 bottle-trays 95 % loss of €160.000,	5340			152,000					152,000
Beer bottles 100% of new purchase to the closed loop system	5341			45,000					45,000
Pallets 2% loss of €14.200,	5350			284					284
Subtotal		**0**	**0**	**240,184**	**0**	**0**	**0**	**0**	**240,184**
1.4. Operating Materials									
Cleaning agents 100%	5400	0	190,000						190,000
Neutralisation agent 100%	5401	0	35,000						35,000
Refrigerants 100%	5402		40,000						40,000
Filtering agents 100%	5403		10,000	10,000					20,000
Laboratory material 100%	5404	0		20,000					20,000
Lubricants 100%	5405	0		11,000					11,000
Tools and maintenance supply	5500			5,000					5,000
Subtotal		**0**	**275,000**	**46,000**	**0**	**0**	**0**	**0**	**321,000**

1.5. Water									
Water from own well (only depreciation and operating materials)			0						0
Water consumption from public supply (hl)	5650		50,000						50,000
Subtotal		**0**	**50,000**	**0**	**0**	**0**	**0**	**0**	**50,000**
1.6. Energy									
Electricity	5600	275,000							275,000
Heating oil 100%	5601	200,000							200,000
Natural gas electricity production, 33% loss of energy efficiency of €21.300	5602	7,100							7,100
Diesel vehicle fleet 100%	5603	200,000							200,000
Subtotal		**682,100**	**0**	**0**	**0**	**0**	**0**	**0**	**682,100**
1.7. Processing costs									
5% loss of beer production	Financial statistics and calculation sheet for production costs		50,000						50,000
Subtotal		**0**	**50,000**	**0**	**0**	**0**	**0**	**0**	**50,000**
Total Category 1		782,100	385,243	514,214	0	0	0	0	1,681,557

(continued)

Table 8.5 (continued)

Environmental costs detail									
Environmental domain	Operating costs (current expenditures)								
Environment-related cost categories	Data source	Air and climate	Water + waste water	Waste	Soil, surface and groundwater	Noise, vibration and odor and fire	Nature protection	General environm. MS	Total
2. END-OF-PIPE									
2.1. Equipment Depreciation									
CC Waste Water treatment									
Waste water treatment plant	Depreciation according to cost center		22,000						22,000
Separating waste water system	Depreciation according to cost center		50,000						50,000
Brewhouse:									0
Dust filter	Estimated depreciation		4,300						4,300
Vapor compaction and control system, 100%	Depreciation according to cost center		7,000						7,000
Hot water recovery, condiment cooling	Depreciation according to cost center		1,000						1,000
CC Fermentation- & storing cellar:									0

Chemical store, 100%	Newly renovated, depreciation estimated on the basis of renovation costs				1,030				1,030
Yeast disposal equipment, also used for recovery of residual beer	Depreciation estimated			30,000					30,000
Pendular gas pipeline	Depreciation estimated		4,800						4,800
CC Carbonic acid system									0
CO_2 recovery and alert system	Depreciation according to cost center	1240		0					1,240
Subtotal		**1,240**	**89,100**	**30,000**	**1,030**	**0**	**0**	**0**	**121,370**
2.2. Operating Materials									
For the equipment defined in Section 2.1 and available on separate cost centre reports, operating materials can be taken from there and deducted from Section 1.4.									0
Operating materials waste water treatment plant	CC 500 (without 5401)		54,500						54,500
Maintenance waste water treatment plant	CC 500		13,700						13,700

(continued)

Table 8.5 (continued)

Environmental costs detail									
Environmental domain	Operating costs (current expenditures)								
Environment-related cost categories	Data source	Air and climate	Water + waste water	Waste	Soil, surface and groundwater	Noise, vibration and odor and fire	Nature protection	General environm. MS	Total
Isolation of steam and water pipes	External services according to cost center	7,000							7,000
Subtotal		**7,000**	**68,200**	**0**	**0**	**0**	**0**	**0**	**75,200**
2.3. Water and Energy									
For the equipment defined in Section 2.1 and available on separate cost centre reports, water and energy can be taken from there and deducted from Sections 1.5 and 1.6.									0
Energy Waste water treatment plant	CC 500	5,000							5,000
Subtotal		**5,000**	**0**	**0**	**0**	**0**	**0**	**0**	**5,000**
2.4. Internal Personnel									
For the equipment defined in Section 2.1 and available on separate cost centre reports, internal personal can be taken from there									0
Personnel waste water treatment plant	CC 500	27,300							27,300
15% of CC Maintenance	CC Maintenance							40,000	40,000

Personnel for waste management	Estimate: 5 people with an average annual person cost of €50,000. 50% of their time			125,000					125,000
Subtotal		**27,300**	**0**	**125,000**	**0**	**0**	**0**	**40,000**	**192,300**
2.5. External Services									
External service for waste disposal	7220 CC HSEQ			10,000	0				10,000
External services for spill management	7220 CC HSEQ				3,000				3,000
External services from lawyers and attorneys for environmental permits	7750 CC HSEQ							500	500
External services for analytical laboratory services	7230 CC HSEQ		500						500
15% of CC Maintenance	CC Maintenance							20,000	20,000
Etc., need to be posted to cost center Environmental Management (EM) or defined by environmental manager, so that the costs can be recorded									0
Subtotal		**0**	**500**	**10,000**	**3,000**	**0**	**0**	**20,500**	**34,000**

(continued)

Table 8.5 (continued)

Environmental costs detail									
Environmental domain	Operating costs (current expenditures)								
Environment-related cost categories	Data source	Air and climate	Water + waste water	Waste	Soil, surface and groundwater	Noise, vibration and odor and fire	Nature protection	General environm. MS	Total
2.6. Fees, Taxes and Permits									
License fee for packaging materials	7100			20,000				0	20,000
Environmental permits	7102		5,000					0	5,000
Waste disposal fees	7105			44,000					44,000
Waste water treatment fees	7106		100,000						100,000
CO_2 allowances bought (EUA, ERU, CER)	7108	0							0
Fee for remediation of disposal dumps	7103			0					0
Other environmental fees and taxes, if applicable								0	0
Subtotal		**0**	**105,000**	**64,000**	**0**	**0**	**0**	**0**	**169,000**
2.7. Insurance									
Environmental part of liability and risk insurance, e.g. for transport of hazardous goods	7700			0	0				0
Subtotal		**0**	**0**	**0**	**0**	**0**	**0**	**0**	**0**

2.8. Remediation and Compensation									
Environmental cost related with remediation and abandonment	7220				0				0
Environmental cost related with compensation to third parties, e.g. farmers and fisheries	7240 CC HSEQ				700		0		700
Biodiversity and landscaping	7670 CC HSEQ				1,000		0		1,000
Subtotal		**0**	**0**	**0**	**1,700**	**0**	**0**	**0**	**1,700**
Total Category 2		**40,540**	**262,800**	**229,000**	**5,730**	**0**	**0**	**60,500**	**598,570**
3. INTEGRATED PREVENTION									
3.1. Equipment depreciation									
Electricity production (block heat with own organic material and power plant), 33% conversion loss, 33% of depreciation of €110.400	Depreciation of fixed assets register	36,800							36,800
Rainwater collection system	Depreciation of fixed assets register		500						500
Bicycle stand and company bicycle	Depreciation of fixed assets register	500							500
Subtotal		**37,300**	**500**	**0**	**0**	**0**	**0**	**0**	**37,800**

(continued)

Table 8.5 (continued)

Environmental costs detail									
Environmental domain	Operating costs (current expenditures)								
Environment-related cost categories	Data source	Air and climate	Water + waste water	Waste	Soil, surface and groundwater	Noise, vibration and odor and fire	Nature protection	General environm. MS	Total
3.2. Operating Materials, Water, Energy									
For the equipment defined in Section 3.1 and available on separate cost centre reports, operating materials, water and energy can be taken from there and deducted from Section 1.4									0
Subtotal		**0**	**0**	**0**	**0**	**0**	**0**	**0**	**0**
3.3. Internal Personnel									
Time to prepare Environmental Impact Assessments and other environment related negotiations and communications of the management board	Estimate: 2 people with annual average person costs of €200,000. 5 % of their time							20,000	20,000
Time of the environmental manager	Estimate: environmental manager (70%) and his substitute (30%), average annual person costs 100.000,							100,000	100,000

Time of the environmental team	Estimate: 10 people 10% of their time, average annual person costs 100.000							100,000	100,000
Other internal personal attending environmental trainings and meetings	Estimate. 100 person hours at average costs of €250							2,500	2,500
For the equipment defined in Section 3.1 and available on separate cost centre reports, internal personnel be taken from there									
Subtotal		**0**	**0**	**0**	**0**	**0**	**0**	**222,500**	**222,500**
3.4. External Services									
Services for Environmental impact assessments and other environmental studies	7760 CC HSEQ							5,000	5,000
External consultants for environmental trainings	7770 CC HSEQ							2,000	2,000
External audit of Environmental Management System	7750 CC HSEQ							3,000	3,000
Ecolabel for returnable bottle	7110 CC Bottling							300	300
Subtotal		**0**	**0**	**0**	**0**	**0**	**0**	**10,300**	**10,300**

(continued)

Table 8.5 (continued)

Environmental costs detail									
Environmental domain	Operating costs (current expenditures)								
Environment-related cost categories	Data source	Air and climate	Water + waste water	Waste	Soil, surface and groundwater	Noise, vibration and odor and fire	Nature protection	General environm. MS	Total
3.5. Other									
Creation, layout and printing of the environmental report	7650 CC HSEQ							5,000	5,000
Subtotal		**0**	**0**	**0**	**0**	**0**	**0**	**5,000**	**5,000**
Total Category 3		**37,300**	**500**	**0**	**0**	**0**	**0**	**237,800**	**275,600**
4. RESEARCH and DEVELOPMENT COSTS									
Pilot project on biodiesel	7760	10,000							10,000
Total Category 4		**10,000**	**0**	**0**	**0**	**0**	**0**	**0**	**10,000**
5. FINES									
Environmental fines	7120				0			0	0
Total Category 5		**0**	**0**	**0**	**0**	**0**	**0**	**0**	**0**
6. LESS TANGIBLE COSTS									
not accounted for									0
Total Category 6		**0**	**0**	**0**	**0**	**0**	**0**	**0**	**0**
TOTAL ENVIRONMENT-RELATED COSTS (1. + 2. + 3. + 4. + 5. + 6.)		869,940	648,543	743,214	5,730	0	0	298,300	2,565,727

7. ENVIRONMENT-RELATED EARNINGS									
7.1. Other Earnings									
Malt dust	4101	0		–500					–500
Yeast sludge	4101	0		–3,000					–3,000
Sale of draff	4100			–35,000					–35,000
Subtotal		0	0	–38,500	0	0	0	0	–38,500
7.2. Subsidies									
Subsidy for research project on biodiesel	4305	–5,000							–5,000
Investment grant for combined block heat combustion, off-set of annual depreciation	4400	–3,000							–3,000
Subtotal		**–3,000**	**0**	**0**	**0**	**0**	**0**	**0**	**–8,000**
TOTAL ENVIRONMENT-RELATED EARNINGS		**–3,000**	**0**	**–38,500**	**0**	**0**	**0**	**0**	**–46,500**
TOTAL ENVIRONMENT-RELATED COSTS & EARNINGS		**866,940**	**648,543**	**704,714**	**5,730**	**0**	**0**	**298,300**	**2,519,227**

Chapter 9
How to Organize an EMA Pilot Project

Chapter 9 describes how to organize an EMA pilot project. The competencies of the project team, selection of sites for pilot testing and a general project plan are discussed. The result of such an EMA pilot assessment may be a company specific adoption of the excel template with more specific cost categories and predefined sources of information as well as an internal procedure which specifies roles and responsibilities. Extracting EMA data from Enterprise Resource Planning Systems and possible elements of an internal EMA assessment standard are explained based on experiences of case studies with Verbundgesellschaft, OMV and Petrom. The chapter ends with a summary of recommendations from about 50 case studies performed so far. The outlook tries to analyze, why companies have been so slow in adopting EMA and MFCA since there is little merit in two separate information systems in an organization, one for financial and cost accounting, the other for process technicians, if "in principle" they should be the same, following the material flows through the company.

9.1 Defining System Boundaries and Sites for Pilot Testing

The input-output-analysis of material flows can be further subdivided from the company and cost centers level to the product produced (Jasch 1998). Product assessments may comprise two system boundaries. Company internal is the attribution of the cost center data to the products produced. The other assessment focus follows the product throughout its life-cycle by adding upstream and downstream life-cycle stages. This method, based on material flow thinking, has been incorporated into the ISO 14040 series for product life cycle assessments.

The input-output balance on corporate level can be calculated on an annual or a monthly basis and is linked to financial and cost accounting, storage and purchase systems. The essential system boundary for corporations is the company fence and the profit and loss accounts. In-depth data are not often available, i.e. balance sheet data for sites within the corporation.

C. Jasch, *Environmental and Material Flow Cost Accounting*: Principles and Procedures, Eco-Efficiency in Industry and Science 25,

Data assessment can be product-, site- or corporation-oriented. Some companies collect and publish for all three levels. Corporate reports are mainly published by multinational companies and contain data, which has to be aggregated from different sites and companies. Often, corporations own shares but not total ownership, of their reporting entities. Thus, questions of consolidation as in financial reporting arise.

Financial accounting and reporting standards which deal with different legal constructs through which corporate control is exercised (e.g. joint ventures, associates or subsidiary operations) should be applied also to internal and external environmental reporting. For the aggregation of EMA data, the following issues may impact on interpretation:

- Establishment or closing of production lines or treatment facilities of the operation
- Acquisition or sale of sites and subsidiaries (and the possible need to adjust prior year data accordingly)
- Outsourcing and it's impact on historic trend data
- Non-adjustment for internal deliveries within consolidated sites

Financial accounting standards have defined three methods of consolidation, depending on the share with which a company participates in another company (Schaltegger 2000 & 1996):

- **Full consolidation** is used by the parent company which controls the majority of the voting rights of a subsidiary (50–100%). The parent overtakes the complete profit and loss account by adding together assets, liabilities, equity, earnings and expenses and deletes all internal deliveries within the group.
- The **equity method** is used for associates, which are neither a subsidiary nor a joint venture to the parent, but in which he has a significant influence (between 20% to 49%). The equity method considers the actual change in value of the share of the equity, but does not integrate sales, assets or liabilities. All internal deliveries are eliminated.
- The **proportionate method** is applied for investments between 1% to 19% of the share capital as well as for joint ventures. Typically, the value of the shares in the books remains unadjusted till significant changes occur.

In environmental reports the degree of ownership of sites is hardly ever mentioned. Also the method of consolidation is hardly ever disclosed or even discussed. In practice, many companies fully consolidate subsidiaries of more then 50% ownership, but without adjustment for internal deliveries, and neglecting minority investments. Thus, the consolidating practices and system boundaries for financial and environmental data assessments can differ significantly. Comparison and relating financial data like turnover and EBIT to environmental data like energy use or total CO^2 emissions may thus be significantly hampered.

Resulting recommendations are

- All sites and subsidiaries should apply the same definitions for data collection.
- All sites and subsidiaries should apply the same input-output chart of accounts for the material flow balance (with site specific details, but aggregating to the same sub-categories).

- Before benchmarking sites, process flow charts must be compared and harmonized.
- All sites and subsidiaries should apply the same consolidation methods.
- The consolidation principles should be disclosed.
- Internal deliveries should be adjusted.
- When calculating key figures, the same consolidating principles should be used as in financial and environmental accounting.

Some corporations with many sites and companies have started internal environmental information systems that collect data from all sites and affiliates and produce corporate environmental reports in addition to site-specific emission monitoring and reporting. Often, international corporations comprise numerous sites and entities, which deliver to subsidiaries and affiliates of the same corporation world-wide.

Adjustment of internal deliveries within plants of a corporation is often only performed for financial data, but not for material flow and other environmental data. Thus, caution has to be paid when relating these figures to each other. If only the inputs and outputs of each site are aggregated without adjustment of supply from within the corporation, there will be numerous double counting. On the other hand, data for turnover and profit will have been adjusted to net values because of financial reporting standards requirements. Thus, the two figures can no longer be related to each other.

For benchmarking projects, it is important to precisely define the process flow chart. Only when the ranges of products—including their packaging—are homogeneous, benchmarking of data will allow useful interpretations. Outsourcing of critical processes, like transport and delivery, cleaning and sanitation, etc. may significantly influence material input and emissions data.

Breweries provide a good example for these effects. For data comparison between production sites there is a significant difference whether, for example, a malting house is a component of the brewery, or whether the brewery acquires its malt from external sources. Similarly, it is of importance for comparison of water and energy data whether bottling occurs on all or only on certain sites and whether all sites bottle in glass and aluminum cans and kegs. In Austria, most breweries also have a non-alcoholic production line for lemonade, which can also distort comparison. Table 8.2 shows the production flow scheme of a brewery.

Still, most corporations and products are more complex than breweries, so the definition of system boundaries has to focus on specific process steps for specified products and defined product life-cycle stages. When comparing companies and products with regard to environmental performance, it is essential that the system boundaries upstream and downstream are identical. But big organizations tend to include most of a product life-cycle stages within their own production plants, while small companies are focused on specific production steps and outsource other production steps.

For performance evaluation and product life cycle assessment (LCA), the production steps and processes covered by the companies or product systems analyzed, must be carefully defined, so that the production steps covered by an input output analysis are identical. Figure 1.2 shows the product life-cycle scheme. Data comparison within sites, processes and products requires that the system boundaries of the participant are comparable; otherwise the results will be meaningless.

9.2 Developing a Project Plan

This section provides instructions for performing a first EMA assessment on site, which consists of a 1–2 days workshop and assesses the total annual environmental and material flow costs of an organization of the previous business year. This top down approach allows planning of measures to improve data quality, perform more detailed assessments on a process or product level and other more detailed surveys, as well as calculating savings and investment projects (Jasch 2006b). The basis for this is always the previous year's costs for a defined system boundary.

In financial accounting, the term expenditure is used. Cost accounting talks about costs, which have slightly different values. Which values are assessed depends on the organization's accounting system. For the first EMA assessment it is recommended to focus on total annual environmental expenditure, which may include calculatory depreciation and interest taken from cost accounting. External costs and future changes in price are not regarded. The assessment is not for calculating investment alternatives, project costs, or potential savings. These can be calculated separately once the annual costs have been assessed.

For the assessment, it has sometimes been practical to split the people involved into separate assessment groups after the general project approach has been agreed upon and a common language established. Some of the assessment steps can thus be prepared simultaneously and are later being jointly discussed, cross checked and agreed upon.

It is recommended to involve the production manager, the environmental manager, the controller, and at least one member from financial or cost accounting. In small organizations, these functions and the related information may be available by only two people. If this is the case, then the assessment groups refer to the timely sequence of the assessment.

In larger corporations representatives from the following departments may be involved:

- Health, safety, security and environment (HSE) – corporate level (project management)
- Health, safety, security, environment and quality (HSEQ) – business segments level and site level for the onsite assessments
- Engineering, planning, production of the business segments and sometimes of the site
- Finance, Cost accounting (site, business segment and sometimes corporate) and partially Asset Management and Warehouse Management

The workshops also help to develop a common understanding on environmentally relevant equipment and cost categories as well as a team spirit for the cost assessments as such. Sometimes the workshops may be structured also as internal trainings on environmental protection and as an awareness raising effort regarding materials efficiency and environmental protection.

The structure of the workshop is recommended as

- Presentation of the project, its goals, terminology and methodology
- Discussion of the system boundaries for data assessment and existing information sources from accounting and other records
- Compilation of the mass balance, discussion of loss and scarp percentages, development of NPO
- Development of a list of environmentally relevant equipment for the business unit assessed
- Collecting data from the related cost centers
- Tracing of the other cost categories
- Cross checking for consistency and completeness with the complete project team
- Recording of open issues and recommendations to facilitate future data assessments

Starting point for an EMA assessment is establishing the mass balance in volumes and recording the related material consumption prices. This often reveals recommendations for stock management, regarding the consistent recording of volumes in stead of other units and regarding the posting of changes in stock to the different specified material categories. Next, the loss percentages for different raw and auxiliary materials have to be agreed upon between the accounting department and production, which may use average standard estimates, and the production and quality managers, who may have additional data estimates and records which are based on actual production experiences.

The next step in the assessment is the definition of the different environmentally relevant equipments, which are separated in end-of pipe technologies and integrated prevention technologies. The environmental shares may have to be estimated by production and the environmental manager. In addition, equipment producing significant amounts of waste and emissions may also be defined. For all these types of equipment the accountants trace or estimate the annual depreciation. Sometimes, the accounting records don't allow any tracing of related postings. Than the production manager may have estimates on former investment costs, and depreciation may simply be estimated with an average 10% as well. The goal is to define a procedure for better recording of related equipment for the future. The goal is not to perform a complete assessment of the past.

For sites operating an environmental management system, the environmental manager should reflect on the projects carried out last year and on any other significant environmentally relevant activities. Tracing the costs related to these activities and the remaining EMA cost categories from the various accounts and previously defined cost centers is the last step for completing the EMA assessment.

The goal of the EMA assessment is to

- Be able to present the total environmental costs of the previous year according to Table 6.1 to the top management and

- To discuss procedures to improve the information systems and technical processes

By using the explanations given, the checklists provided in the annex and the Excel template it should be able to assess the environmental costs of the previous year in 1 to 2 days.

Experience from several case studies has shown that while the structure for the workshop was always identical, the time spent for specific topics varies significantly and depends totally on the availability of data in the existing information systems. It is recommended to focus on developing of recommendations for the improvement of information systems and rather work with rough estimates than to spend too much time for tracing outdated data in inadequate information systems.

The cost assessment reveals improvement options in two areas:

1. What always can be found, are options and measures necessary to improve the quality and consistency of data and information flows in an organization. This is the starting point of most projects and the focus of most follow up projects.
2. In companies, that have not done environmental management projects for several years, also technical improvement options may become obvious. What always is made visible, mostly for the first time, are the costs related to inefficient production, wasting materials and energy. So even if the technical solution might not be known at the end of the first assessment, the priority areas for deeper investigation will have been defined.

Other results of the assessments may include a changed focus of what are significant costs and options for improvement as well as a better awareness of the assessment team of the total material flow and environmental management related corporate cost structure.

9.3 Extracting EMA Data from Enterprise Resource Planning Systems

The Verbund group is Austria's largest producer and transporter of electricity, generating about 50% of the electricity consumed in the country. It is one of the leading hydropower producers and also one of the most profitable energy utilities in Europe. With approx. 2,400 employees, Verbund generates annual sales of more than 3 billion Euros. The group consists of the corporate parent and a number of subsidiaries (energy generating companies, a grid operating company, etc.) (Verbund Sustainability Report 2003). In 1994, Verbund started to report on its performance on environmental issues, including some environment-related costs for measures taken to avoid or minimize environmental impacts.

In 2001, Verbund decided to take part in a pilot project that would assist the company to better assess environmental performance and environment-related costs via more rigorous EMA (Jasch and Schnitzer, 2002). Three different sites, each representing one of Verbund's business groups, were chosen to take part in the pilot project: a small hydro power station, a fossil fuel power plant and a substation of the transmission grid. At each site, an assessment of annual costs was performed, and intensive discussions were held as to which costs would be defined as environment-related. Agreement was reached that costs driven by environmental regulation or community concerns about environmental issues would be defined as environment related.

The Verbund had reached an agreement with the Austrian Environmental Protection Agency regarding the environmental domains to be considered. They are specified for each of the business groups (hydropower, thermal power and the grid). Within the EMA project the domains for each business group were streamlined, made compatible and the terminology was made consistent. A column for environmental management was introduced for all business groups. The EMA excel templates were adapted to the new structure.

The EMA assessments revealed, that before the EMA project environment related costs had only been collected from sites and related to equipment and maintenance, but that significant cost also occurred in administrative departments, e.g. related to projects and research and development or at head quarter, e.g. the costs for compensation to fisheries and farmers.

It was also necessary to clearly specify which data would be needed from the company's Enterprise Resource Planning accounting system (from SAP), where to find it, define responsibilities and avoid double counting. Within the SAP system, environment-related costs can be found in two different places:

1. Data records associated with a specific company project or
2. Cost center data records

1. Data records associated with specific company projects

The procedure in Fig. 9.1 shows how environment related costs are collected from company projects. Starting point are the action plan reports generated from SAP. They in detail categorize the related types of costs, e.g. like

0F.xxxxx operating expenses for regular measures
0H.xxxxx maintenance for regular measures
0S.xxxxx other projects realized with regular measures
1B.xxxxx Investments in operating equipment
1H.xxxxx Maintenance
1S.xxxxx Other projects
1E.xxxxx Investments beyond current performance
1P.xxxxx prestudies
1K.xxxxx customer projects
1V.xxxxx insurance projects

The types of costs possibly relevant for EMA were specified in detail. For projects qualified as environmentally relevant at the point of time of project approval, the environmental share is established annually by the environmental manager in consultation with the project leader. A consistent and comprehensible approach over the years is recommended.

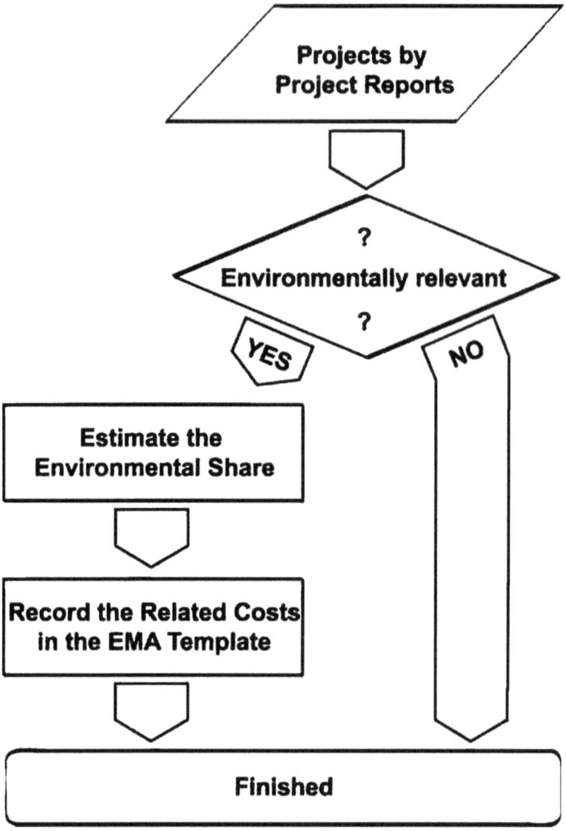

Fig. 9.1 Data records associated with a specific company project

The project related environmental costs are inputted by the environmental manager into the EMA excel templates, which have been specifically adapted to the needs of the business groups.

2. Cost center data records

The procedure in Fig. 9.2 shows how environment related costs are collected from company cost center reports. Starting point are the list of accounts as well as the cost center reports generated from SAP for each site. It is specified, that only the cost categories 4xxxxx till 7xxxxx are to be considered, which are directly related to expense accounts, but not the cost categories 9xxxxx which contain the allocation of overhead to cost centers.

Analogous to the cost assessment for projects the environmental relevance of the cost center is determined, the appropriate share estimated and the different postings are collected in the EMA excel template, which already contains predefined fields for the cost centers, which in general are relevant for EMA.

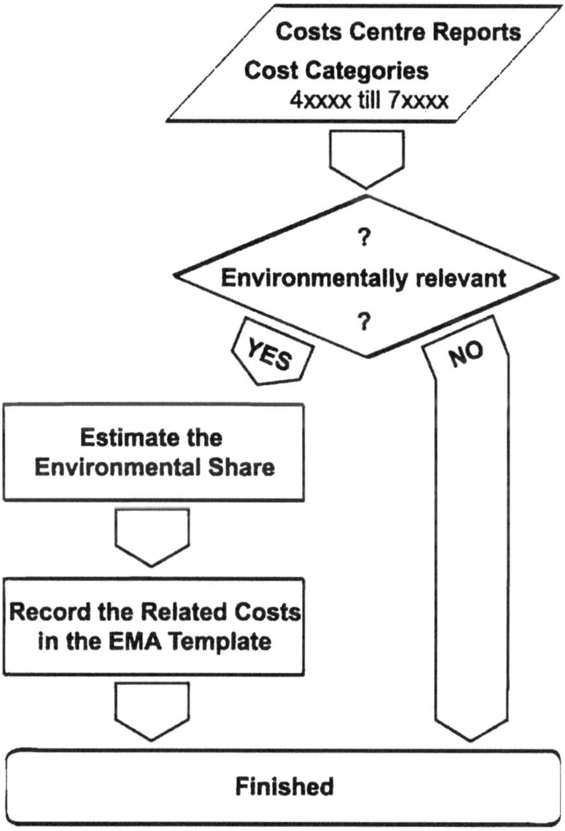

Fig. 9.2 Tracing environmental costs from cost center data records

The company then had to determine which costs, taken from both data records, were environment related, per the company's definition. For costs with dual characteristics, for example, those considered being both environment and efficiency related; an appropriate percentage of the cost was taken to be environment related.

During the course of the pilot study, it became evident that it was not possible to have the SAP software automatically extract and report the needed environment-related cost data. So, a formal data collection procedure was written for environmental managers from the about 150 sites of the three largest subsidiary companies. This procedure helps the managers to extract SAP data that must be reported to the corporate parent company each year. The extracted data are reported by cost category and by environmental domain. Environment-related earnings are also reported.

In 2003, all of Verbund's power generation companies and its grid operating company adopted this EMA process developed during the pilot project. The data is being used for internal management decision making and external reporting at both the corporate and site level, and allows performance benchmarking between different sites.

9.4 Elements of an Internal EMA Standard

Larger corporations with several sites like Danisco, OMV, Petrom and Verbundgesellschaft have all implemented an internal data collection procedure as one of the outcomes of their EMA pilot assessments, which in detail specify definitions, responsibilities and data sources for the corporate EMA assessments. Harmonized and auditable data quality for longer time periods and from several sites is thus secured.

This chapter develops a possible outline of such a procedure and is primarily based on the experiences gained with implementing the system at OMV, Petrom and Verbundgesellschaft.

OMV is the leading oil and gas group in Central Europe. Its business activities cover every stage of oil and gas production, processing and marketing, as well as petrochemicals. With sales revenues of EUR 20 billion in 2007, 33,665 employees and a market capitalization of around EUR 15 billion at year-end 2007, OMV Aktiengesellschaft is the biggest listed industrial enterprise in Austria.

OMV is a rapidly expanding company. At the end of 2004, OMV acquired a 51% stake in Petrom, which was previously a state-owned company in Romania. At the beginning of 2005, an extensive HSE integration program was drawn up to introduce international best practice models to health, safety and environmental standards of Petrom. OMV and the Romanian government have agreed upon modalities to deal with contamination due to operations prior to 2005. Therefore, systematic information about environmentally relevant projects and cost is needed not only from a managerial point of view but also with regard to contractual terms.

In addition, Petrom reviewed corporate wide standards for consolidation of financial accounts, which were implemented by 2007. Therefore requirements regarding the information system such as those related to the implementation of a standardized method for reporting environmental costs were defined still in autumn 2006.

Monitoring and measuring Health, Safety and Environmental (HSE) performance requires reporting procedures that are able to collect data on these various organizational levels and help consolidating them up to the Group level. OMV applies an information tool with web-based user interfaces that facilitates simple and standardized data input and validation procedures as well as group wide data consolidation and reporting. The tool is flexible enough to map organizational changes and to adapt the indicator list according to specific reporting needs. HSE indicators for the following topics are reported regularly (monthly or annually):

- Organizational information
- HSE management (including indicators for environmental costs)
- HSE events and highlights
- Occupational health
- Safety
- Energy

- Greenhouse gases
- Environment

Several types of HSE reports are based on this information:

- Internal monthly reports
- Internal annual HSE report with detailed evaluation of key performance indicators
- Annual update of HSE highlights and figures for external communication
- Corporate HSE Report every 2 years, which is published on the corporate internet site

Furthermore, environmental performance, investments and annual costs have to be reported to several agencies:

- National statistical agencies assess the annual investments for environmental protection as well as annual costs.
- National environmental protection agencies assess environmental impacts.
- Local authorities require information on environmental protection equipment.
- The Romanian government requires information regarding clean up of contamination and mandatory investments to fulfill environmental permits.
- The corporate financial management report from 2005 onwards also has to disclose key non financial performance data regarding environmental protection.

It was therefore in the core interest of OMV to base these disclosure requirements on sound definitions and auditable data quality. The HSE monitoring system was linked to existing information systems and thus allows efficient and consistent data reporting.

The project goals were to develop and establish a Group-wide consistent methodology for Environmental Management Accounting based on internationally recognized standards in order to

- Establish sound data about environmental costs for decision making
- Support reporting to authorities and agencies
- Facilitate internal and external benchmarking of environmental costs
- Facilitate data and reports for internal and external communication of environment-related costs

The main goal of the project was to develop better definitions and procedures for the assessment of the costs for environmental protection (mostly end-of- pipe technologies) and costs for integrated prevention, that are specifically adopted to the situation and requirements within OMV and Petrom. The focus of the project was not so much on assessment of detailed data for past years, but on developing in the course of a pilot assessment definitions and procedures for the future based on the experiences from data shortcomings. However, re-assessment of the cost for the year 2005 was taken as a starting point for this exercise.

One of the challenges of this project was to map the technical and financial information sources and ensure data consistency and completeness. The financial systems are structured by subsidiary companies and profit centers and follow the logic of financial markets and tax law, while the technical information systems are oriented towards production processes, sites and equipment; and the reporting tool is set up for internal and external HSE reporting purposes.

The data were collected from the following information sources:

- Financial accounting and the list of accounts
- Cost calculation reports
- Cost center reports and profit center reports
- Asset management
- Inventory statistics
- Production planning
- Technical monitoring systems
- Recording of personal hours spent
- Recordings of quality management
- Waste statistics
- Environmental report
- Etc.

When collecting environmental data from the sites there is a tendency to overlook environmental costs of the more service oriented administrative departments. For instance research activities which are environmentally relevant are not necessarily included in standard HSE reporting, as they may be carried out by other departments and expenditures are directly posted to service cost centers and not related to the physical flows of a specific reporting unit.

With regard to the development and implementation of the EMA methodology major attention was drawn during the project to

- Organizational aspects of conducting EMA assessments jointly between finance, controlling, production and HSEQ staff with an interdisciplinary approach
- Definition of system boundaries for EMA assessments in the different business segments of OMV Group and
- Identification of data resources and information gaps

Once the methodology was established, reporting of environmental costs was integrated in the annual HSE data collection and reporting as of the HSE 2006 Data Collection Campaign. Business divisions, hence, started implementation of EMA by using the 2005 pilot assessments carried out during the project as templates for 2006 reporting.

The IFAC approach for EMA has been adjusted to the specific requirements in the oil and gas industry and especially OMV and Petrom. The system boundaries and assessment units for physical and monetary data were defined.

The definition and listing of environmentally relevant equipment for the business units ensures that all business units in all countries where OMV is present

1. Recognize and record the same equipment as environmentally relevant
2. Allocate them to the same environmental media
3. Consistently classify them as end-of-pipe or prevention technology

In addition, the project has resulted in increased and consistent understanding about what is environmentally relevant and should be reported. The data quality for consolidated reports has thus increased significantly, likewise the comparability of data for internal benchmarking, as data and data sources are traceable and harmonized.

The roles at data collection and the necessity to work in interdisciplinary teams as well as the responsibilities for the several data recordings were clarified. There is consensus that EMA cannot be done solely by the environmental manager without the input of the financial and production department, while at the same time cost accounting cannot produce the data without the environmental department.

The assessments have led to recommendations regarding new accounts, cost centers, accounting procedures and classifications in the information systems that were still implemented in the year 2006, to facilitate data assessment for future years.

It can be assumed that environmental data collection from the year 2007 onwards will be significantly less time consuming as the EMA excel spreadsheets developed provide a consistent structure, that serves several internal and external reporting requirements.

A possible outline of an internal EMA standard could be structured like

1. Objectives and scope
2. Definitions
3. Responsibilities
4. Material flow data (input-output analysis)
5. Environmental cost categories and assessments
6. Procedure for data gathering
7. Internal reporting
8. External reporting
9. Appendix: Excel templates for EMA assessment by business groups

And contain the information described in the following.

9.4.1 Objectives and Scope

The objective of the EMA assessment is to provide a sound data basis for internal management and to meet internal and external reporting requirements with accurate and reliable data.

The standard should define the reporting scope and system boundaries for data assessment. It is recommended to apply the standard to all fully consolidated subsidiaries and holdings not fully consolidated, where the corporation has a controlling interest. It is thus consistent with the financial reporting system boundaries.

The main focus of the standard may be on monetary information. It should also define links to environmental data management and to purchase, warehouse management and stock keeping with regard to physical data collection.

9.4.2 *Definitions*

This section should provide the definitions for the distinction between end-of-pipe treatment and integrated prevention. In addition, the environmental cost categories and environmental media may be defined here. In may be helpful to also quote conversion factors.

9.4.3 *Responsibilities*

Since an EMA assessment requires information and data input regarding physical and monetary information, an interdisciplinary approach is necessary in order to achieve completeness in the information gathering process.

The main responsibility for the timely performance and reporting of the annual EMA assessment may be with the environmental manager, but additional persons need to be involved. For a large corporation like OMV, the following setting seems appropriate. For smaller organizations, coordination between the environmental manager, controlling and process monitoring needs to be insured. The reporting cycle should be identical to financial reporting. Some organizations with financial reports not at the end of the calendar year might encounter the necessity to install two reporting cycles, for the financial report at the specified time and in addition at the calendar year for several other reporting requirements.

The environmental manager may be responsible for

- Coordinating the annual EMA assessment as part of the annual internal reporting taking place between January and March every year; this includes the appropriate coordination and consultation with accounting, controlling and technical staff required for the assessment.
- Checking the completeness of environmental costs, especially regarding the inclusion of new projects and investments (annual screening in coordination with accounting and controlling staff).
- Checking the plausibility of environmental costs and environmentally relevant cost portions of both annual expenditures and investments.
- Completing the figures of environmental costs in the annual environmental data collection according to the defined environmental performance indicators.
- Promote appropriate information flows regarding environmental relevancy of investments including project managers and financial staff, and raise awareness for appropriate flagging in the project system and in the fixed asset accounting.

The financial department may be responsible for

- Flagging the environmental-relevant investments in the fixed asset accounting once a new project is entered. The information about environmental relevancy is taken from the project description. Where the environmental relevancy is not explicitly stated in the project description, the responsible project manager shall be contacted for clarification.

- Producing a report containing environmental-relevant investments of new equipments and plants commissioned in the reporting year, plus annual depreciation of all environmental-relevant equipment (optional, where depreciation is required).
- Identification of accounts, cost centers and reports necessary for the EMA assessment.
- Producing reports out of the corresponding systems (accounting, inventories, etc.).
- Collaborating in the discussion of environmental costs during the EMA assessment and the allocation to the corresponding cost categories.

Specific information required for the assessments should be provided upon request by

- (Process) engineers: input and information about processes; environmental relevancy of equipment and installations, etc.
- Planning, production and other departments according to needs
- Research and development
- Project managers: information on environmental relevancy of investments in the project summary in order to facilitate appropriate flagging in the project system and in the fixed asset accounting

The corporate HSE department may be responsible for

- Group-wide consolidation of environmental costs
- Focal point for reporting environmental costs to external stakeholders in coordination with other Headquarter departments (such as: Investor Relations for the Annual Report; the Sustainability Report, reports to statistic agencies, etc.)

9.4.4 Material Flow Data (Input-Output Analysis)

This section should define the degree to which material flow data in physical terms is to be collected, including data sources and responsibilities. Some organizations have very developed systems for physical mass balancing and production monitoring that include all major equipment and can be aggregated and segregated as needed.

On the other hand, it may be difficult to match this mass balance with accounting data as the structure of the cost centers may be quite different to the organization of the physical mass balance units.

Other organizations may decide to only partly collect material input data as part of their EMA assessment. The links to environmental monitoring systems in place should be defined.

9.4.5 Environmental Cost Categories and Assessments

This section should describe the structure of the EMA assessment templates and provide company specific examples for the environmental cost categories and related data sources. In the EMA assessment template the column for source of information is provided therefore and may be pre-filled from the pilot assessments. In order to ensure consistency of data also for previous years for each subcategory of material input in physical and monetary terms as well as for all other cost subcategories the source of information (department, cost center, account, material number, name and number of other technical reports) is being recorded. This will facilitate data tracing in the future and ensure consistency and completeness of the reported figures.

Typical environmentally relevant cost categories of the business groups of relevance should be classified according to international EMA guidelines and the definitions of statistic agencies. Specific guidance with criteria is needed regarding the question, if an equipment is environmentally relevant and to which degree and who is in charge of taking this decision.

It is recommended to carry out the assessments in standardized EMA assessment templates, which are provided in the annex of the internal standard.

The templates typically consist of four sheets:

- Mass balance (I-O balance)
- Environmental costs detail
- Environmental costs summary and
- Structure (environmental costs summary %)

After the completion of the mass balance in physical terms and the recording of the values of materials used, information is only added into the Detail sheet. The cost subcategories are already set.

In order to facilitate traceability and audit ability of EMA assessments, information about data sources (e.g. account numbers, reports, etc.) should be recorded carefully in the corresponding column of the assessment sheet.

Thresholds for the recording and reporting of environmental costs may be set by the reporting units considering the insignificancy of certain costs (e.g. small maintenance bills). If threshold are set, they shall be documented explicitly in the assessment report.

9.4.6 *Procedure for Data Gathering*

This section should give guidance on the sources of information to be used for the data assessment. The following information sources should be considered for the information gathering process:

- Financial accounting and the list of accounts
- Cost calculation reports
- Cost center reports and profit center reports
- Asset management
- Inventory statistics

- Production planning
- Technical monitoring systems
- Recording of personal hours spent
- Recordings of quality management
- Waste statistics
- Environmental report
- Etc.

For the predefined cost categories, these information sources may be already listed. The case study of Verbund shows how the costs are specifically collected from project reports and defined cost centers and accounts. Caution needs to be taken to avoid double counting as well as omission of significant (new) costs, e.g. cost for greenhouse gas monitoring and permits, which may be collected outside the cost center for environmental management.

Clarification should be provided regarding the recording of depreciation. Ideally, the environmentally relevant equipment shall be flagged by the types of environmentally relevant investment categories in SAP or other accounting systems at the point of time when a project code is being defined. Once the SAP flag for the environmental investments is realized, the actual depreciation can be traced by a SAP run and also the total annual amount of environment related investments is available without any further investigation. If this is not possible (or for existing equipment), it may also be a solution to estimate the average life time and not record the actual depreciation from asset management.

The standard may record typical cost centers that need to be investigated for the EMA assessment, e.g.

- Waste disposal dumps (in the case of existing or planned own waste disposal dumps, but not, if waste management is basically outsourced and no equipment and land is used)
- Waste water treatment plants (especially if related with own personal, significant maintenance and chemicals consumption)
- General environmental management

It should be defined if only the costs for the last business year are to be collected and reported or if in addition budgeted data is to be reported as well.

Finally, thresholds for cost recording may also be quoted. For some businesses, the separate recording of operating material costs, water and energy costs as well as personal costs directly related to the environmentally relevant equipments listed above is not always easily available from cost center reports and may be omitted.

9.4.7 Internal Reporting

The EMA assessments are typically carried out once a year in order to prepare the information required for internal management in the first quarter of each year. It is strongly recommended to work in interdisciplinary teams with the participation of the environmental manager, engineers (especially for the determination of environmental relevancy of equipments for integrated prevention) and the financial department (accounting and/or controlling staff).

The use of business specific EMA templates (refer to appendix of the internal standard) is recommended in order to meet the minimum requirements of this standard and to facilitate traceability of assessments for audits.

If some environmental costs cannot be identified with reasonable effort in the accounting and controlling systems, best estimates should be reported and appropriately commented and documented. The reporting format to top management should be defined in the standard.

9.4.8 External Reporting

The minimum reporting standard for environmental costs in order to meet information requests of statistic agencies as well as the standards set by the Global Reporting Initiative (GRI) requires environmental investments and expenditures by type (end-of-pipe vs. integrated prevention) and affected environmental domain.

National statistical agencies require a distinction of environmental protection measures into emission control or prevention and in addition a classification into the environmental media (Air, Ground, Water, etc.) affected. In certain countries a division of the future investments into mandatory or voluntary may also important in order to be able to fulfill reporting requirements to authorities.

9.4.9 Appendix: Excel Templates for EMA Assessment by Business Groups

The annex of the internal EMA standard should provide the EMA assessment templates in Excel format, specifically adapted to the typical cost categories of the business units involved.

It is also important to mention, that EMA data can be mostly collected from accounting records but it still needs recording into a separate file, as much of this data is used for different reporting purposes. The time of internal personal spent for trainings on environmental protection may for instance be significant and can be estimated, but the related costs for personal would still be recorded under the traditional accounts and cost centers.

Often, different business units and countries may have different approaches on what equipment to consider as environmentally relevant and whether to post it under treatment or prevention. For each business unit a checklist with the different types of equipment may therefore be developed based on the checklist provided in the Annex of this book for each environmental domain and attached to the corporate wide internal standard.

9.5 Summary of Recommendations from Case Studies

In the last years several pilot projects have been performed (Jasch Danse 2005, Jasch Lavicka, 2006, Jasch Schnitzer 2002). The case studies resulted in some recommendations, which are applicable in most companies.

9.5.1 Data Collection of Material Purchase by Material Groups in Financial Accounting

In some enterprises the entire material purchase is booked on one account only and it is only possible to evaluate by hand the extensive cost center accounts or stocktaking lists to divide the actual material use into the material groups. As an aid, the recordings of the production manager of materials inputs may be multiplied with average prices, in order to at least be able to indicate orders of magnitude. The fact that such a system cannot strengthen cost consciousness in handling raw, auxiliary and operating materials is obvious.

A clear distinction between the accounts for raw, auxiliary and operating materials is necessary, especially when non-product output (NPO) costs are intended to be assessed. Raw and auxiliary materials are part of the product, thus loss percentages need to be calculated or estimated. Operating materials are by definition not part of the product and thus must become part of waste and emissions. The amounts and values used are often not consistently recorded.

The posting of inventory changes should be carried out separately for the different materials accounts and include a separate recording of the price and volume difference. This way accurate data on materials inputs and outputs in volume and price can be obtained so that the total amounts and values of materials used are available for further controlling measures. Posting of the total difference of inventory change to one separate account leads to ignorance regarding actual materials used.

It should be clearly defined, which material numbers belong to which material group and account. The material groups should be traceable, e.g. by separate accounts.

Volumes should be added gradually to the recordings of material numbers in stock management. This way, consumption would be aggregated automatically into volumes. Consistent use of units (kg) in the ERP system ensures that the total sum automatically aggregated does not have to be manually corrected.

Materials and supplies for maintenance from maintenance services should be recorded separately allowing for the total materials input to be calculated.

9.5.2 Estimation and Recalculation of Material Scrap Percentages

The loss percentages for raw materials, packing material, auxiliary materials and the final product are often based on outdated estimated values and only are recalculated for a few material groups. The employees on-site usually have more precise estimated values than the accountants. A correct recalculation mostly raises frightening results.

It is recommended to check for consistency of system boundaries for material flow accounting in technical and accounting information systems and define, which accounts, cost centers and cost categories must be consistent by amount and value.

The input-output material balance collected by the environmental department and sometimes disclosed in an environmental statement is hardly ever consistent with the system boundaries of the accounts and cost center reports. As a consequence, the data can not be audited for consistency. It is common the for the recording of the costs and amounts of waste several different values and records can be found on one site (record of the environmental manager without the costs for weighting, transport and rent of disposal cans, the financial account with some wrong postings and the accounts of several suppliers with additional services).

9.5.3 Depreciation of Projects/Investments Before the First Year of Cost Assessment

During the first cost assessment, the question is often posed how to deal with missing values of the previous years. If these can be estimated or assessed easily, it should be done. But, the main goal of the first assessment is to improve the data basis for the next years and not detailed and cumbersome assessment of previous values.

A clear corporate and sector specific definition of what is environmentally relevant equipment can be included in the internal standard. When a company has several sites in more than one country often the range of interpretations of what is environmentally relevant of the people carrying out the EMA assessment are broad and often contain highly efficient production equipment as well as maintenance expenses, while others report only end-of-pipe treatment equipment. An interpretation of aggregated data on corporate level is thus hampered.

For the existing equipment, every assessment unit should define the significant environmentally relevant assets for each operating area and try to estimate the investment costs. The annual depreciation may be simply calculated with 10% if the actual expenditure if it is not easily available.

For future investments, the environmentally relevant equipment should be flagged by the types of environmentally relevant investment categories in SAP at the point of time when a project code is being defined. Once the flag for the environmental investments is realized in asset accounting, the actual depreciation can be traced by a system run and also the total annual amount of environment related investments is available without any further investigation.

9.5.4 Distinction to Health and Safety and Risk Management

Designing a system appropriate for the company involved is the most important target. Some companies have added a column for safety and risk prevention, as this duty is also part of the job description of the environmental manager. Health is mostly the duty of other departments. But some companies simply included a new column for "Health and Safety" in the EMA excel assessment template, as this may be covered by the same department.

9.5.5 Product Oriented Pollution Prevention

Companies with significant activities and costs regarding the prevention of waste and emissions of products, e.g. engaged in ecodesign, or developing substitutes for hazardous product components, may decide to include a column for product oriented prevention.

9.5.6 New Cost Centers and Accounts

The creation of new cost centers is recommended for

- Waste disposal dumps (in the case of existing or planned own waste disposal dumps, but not, if waste management is basically outsourced and no equipment and land is used)
- Waste water treatment plants (especially if related with own personal, significant maintenance and chemicals consumption)
- General environmental management

Separate accounts should be established for the different raw, auxiliary, packaging and operating materials. In the list of accounts a distinction should be made between raw and auxiliary materials as well as packaging, which becomes a product with loss percentages. As by definition operating materials are not included in the product, these are converted into waste and emissions.

Accounts for materials and utilities should be clearly distinguished from accounts for services. If only materials are collected on an account than the volumes used may be estimated dividing with average prices. Materials and supplies for maintenance from maintenance services could be separated allowing for the total materials input to be calculated.

Separate accounts for the utilities (energy, water) should be established, defined as direct costs of production.

Earnings from sales of scrap metals; steam condensate etc. should not be offset directly against the materials purchase account. Instead separate accounts for other earnings from by-products should be established.

9.6 Outlook

Since the mid 1980s, several forces have encouraged the shift to prevention-oriented strategies, including public concerns with environmental degradation worldwide and climate change threats, increasingly stringent pollution control requirements in Europe, and widely publicized industrial accidents. As a result, firms have faced a rising tide of public demands for shifts to cleaner technologies and environmentally sound products (Jasch, 2006a).

However, companies have been slow to move away from traditional end-of-pipe strategies toward more prevention-oriented practices. If, as many argue, pollution prevention pays, what accounts for this slow pace of change? If investments in pollution prevention are, in fact, in the interest of the firm, what accounts for the continuing reluctance to move towards a more preventative pollution management mode? And why, in light of the publicized benefits of pollution prevention, do organizations continue to be surprised when prevention-oriented projects produce financial pay-backs to the organization far beyond those expected of many conventional compliance-driven capital investments?

The following explanations for this apparent contradiction may be reasonable:

- The organizational structure and behavior of companies hinders pollution prevention projects from entering the decision-making process, thereby precluding these alternatives from consideration by the companies.
- Barriers linked to the methods of cost accounting and capital budgeting result in a poor visibility of the costs of non product outputs. Even if a pollution prevention project successfully entered the capital budgeting process, competition with other projects for limited capital resources is hampered by the poor knowledge of the true costs of non-product output.
- Psychological and social effects might need consideration when changing information systems. Often, increased responsibility for material flows and altered purchasing and stock management rules are not in the interest of department managers as they may be linked with reduced spheres of power and increased control.

Overcoming the barriers of traditional accounting regarding environmental costs and material flow management have been the focus of this book. In the light of increasing energy and material prices and possible shortages as well as climate change issues these concepts will gain even more importance as they are entering the risk management agenda.

Effective cost accounting requires effective material flow accounting. Environmental costs arise when materials are used, processed and released as non-product outputs. Understanding material flows as they move through a production system is a prerequisite to identifying and tracking environmental costs. Material flow balances are the most rigorous basis for developing such information, but short of this, improved materials accounting and screening process flow diagrams may well be sufficient in the first round. The effects of hiding environmental and material flow costs in overhead accounts and not correctly posting them to cost carriers have been highlighted. The omission of defining and monitoring relevant materials or energy flows can create major cost consequences that may lead to misguided management decision-making.

Improvements can't be achieved by simply installing new software. There is no separate software for EMA that solves all problems. Those seeking such a definitive, all-encompassing stand-alone solution are likely to be disappointed. As environmental and material flow cost information serves different functions and reporting requirements in an organization, EMA is better thought of as a set of adjustments to current cost accounting systems, all with the purpose of identifying, tracking, and reporting environmental and material flow information to sharpen management decisions. More rigorous process flow information, linked with allocation of overhead costs to the respective cost centers and objects is vital. This amounts to nothing more than sound management and engineering practices being applied to cost accounting.

Financial statement audits are increasingly considering general risks. Financial statement auditors seek to understand all significant aspects of business risk facing an organization and how those risks are managed, so as to develop the most effective approach to gain assurance about the reliability of management information and hence of reported information. Business risk can be defined as any probability that the organization will not achieve its business objectives. Accordingly, as sustainability becomes more important to the objectives of a business and hence to its risk management and control processes and in the light of sharply rising energy and material prices, top management and financial statement auditors are increasingly interested as well.

Sustainability reports are increasingly being integrated into financial reports and externally verified. Thus, the disclosure of reliable environmental performance and costs data for the corporation, based on a solid information system that consistently collects and aggregates financial and physical data, is vital. There also is a trend from separate financial and sustainability reporting towards integrated reports. Likewise, there is little merit in two separate information systems in an organization, one for financial and cost accounting, the other for process technicians, if "in principle" they should be the same, following the material flows through the company.

References

Association of German Engineers, VDI 3800, Determination of Costs for Industrial Environmental Protection Measures, Berlin, 2001, Association of Engineers.

Bebbington, J., R. Gray, C. Hibbitt and E. Kirk, Full Cost Accounting: An Agenda for Action. London: The Association of Chartered Certified Accountants, 2001; http://www.accaglobal.com/pdfs/research/ACCA-rr73-001?session = fffffffeffffffffc28288ca4033b4389c11b14f4c8dda97fb1921a6792820c2.

Bennett, M., and P. James. (Eds.), The Green Bottom Line, Environmental Accounting for Management. Sheffield, UK: Greenleaf Publishing, 1998; http://www.greenleaf-publishing.com/pdfs/gblch1.pdf.

Burritt, R., T. Hahn and S. Schaltegger. "Towards a Comprehensive Framework for Environmental Management Accounting – Links between Business Actors and Environmental Management Accounting Tools." Australian Accounting Review, July 2002; http://www.uni-lueneburg.de/eman/pdf_dateien/Burritt-Hahn.pdf.

Canadian Institute of Chartered Accountants, Full Cost Accounting from an Environmental Perspective. Toronto, Canada, 1997.

Environment Canada. Introductory Guide to Environmental Accounting: Environment and Decision-making: An Appropriate Accounting. Ottawa, Ontario: Environment Canada, 1997; http://lavoieverte.qc.ec.gc.ca/dpe/Anglais/dpe_main_en.asp?prev_comp.

Environmental Protection Agency of Baden-Würthemberg. Corporate Material on energy Flow Management, Improving Eco-Efficiency via Sustainable Reorganisation. Karlsruhe, 1999 (available in German only).

Envirowise, Increase Your Profits with Environmental Management Accounting. Oxfordshire, UK, 2003; http://www.envirowise.co.uk/envirowisev3.nsf/key/CROD5HYLHS.

European Commission, Commission Recommendation of 30 May 2001 on the Recognition, Measurement and Disclosure of Environmental Issues in the Annual Accounts and Annual Reports of Companies. Official Journal of the European Union L 156/33, June 13, 2001a.

European Commission, Commission Regulation 761/2001 on Environmental Management and Audit System, EMAS, Brussels, 2001b.

European Commission, Commission Regulation (EC) No 1670/2003 of 1 September 2003 Implementing Council Regulation (EC, Euroatom) No 58/97 with regard to the Definitions of Characteristics for Structural Business Statistics and Amending Regulation (EC) No 2700/98 Concerning the Definitions of Characteristics for structural Business Statistics. Official Journal of the European Union L 244/74, September 9, 2003.

European Commission, Commission Regulation (EC) No 1907/2006 of the European Parliament and of the Council on the Registration, Evaluation, Authorisation and Restriction of Chemicals (REACH), June 1, 2007.

European Parliament and Council, Directive 2003/51/EC of the European Parliament and of the Council of 18 June 2003 on the Annual and Consolidated Accounts of Certain Types of

Companies, Banks and other Financial Institutions and Insurance Undertaking. Official Journal of the European Union L 178/16, July 17, 2003.

Eurostat, SERIEE–Environmental Protection Expenditure Accounts–A Compilation Guide, 2002. European communities, Luxembourg.

Fichter, K., T. Loew and E. Seidel. Betriebliche Umweltkostenrechung. Berlin: Springer, 1997 (available only in German).

Fichter K., T. Loew, C. Redmann and M. Strobel, Flusskostenmanagement, Kostensenkung und Öko-Effizienz durch eine Materialflußorientierung in der Kostenrechnung. *Wiesbaden, Hessisches Ministerium für Wirtschaft, Verkehr, und Landesentwicklung*, 1999 (available only in German).

German Environmental Protection Agency/German Environment Ministry, Eco-Controlling Manual. Munich: Vahlen, 1995 (available only in German).

German Federal Ministry for Environment/Federal Environment Agency, Guide to Corporate Environmental Cost Management. *Berlin*, 2003.

Global Reporting Initiative (GRI), Sustainability Reporting Guidelines on Economic, Environmental and Social Performance. Amsterdam, 2006; http://www.globalreporting.org.

Gray, R. and J. Bebbington. Accounting for the Environment, London: Sage Publications, 2001; http://www.sagepub.co.uk/book.aspx?pid = 101898.

Howes, R. Environmental Cost Accounting: An Introduction and Practical Guide. London: The Chartered Institute of Management Accountants, 2002.

Institute of Chartered Accountants in England and Wales, Environment Steering Group. Environmental Issues in Financial Reporting. London, 1996.

Institute of Chartered Accountants in England and Wales, Information for Better Markets, Sustainability: the Role of Accountants. London, 2004.

International Federation of Accountants, IFAC. Management Accounting Concepts. New York, 1998.

International Federation of Accountants, IFAC. International Guidance Document: Environmental Management Accounting. New York, 2005.

International Standardization Organization, Environmental Management – Environmental Management Systems – Specification, ISO 14001. Geneva, 1996.

International Standardization Organization, Environmental Management – Life Cycle Assessment – Principles and Frameworks, ISO 14040. Geneva, 1998.

International Standardization Organization, Environmental Management – Environmental Performance Evaluation – Guidelines, ISO 14031. Geneva, 2000.

International Standardization Organization, Environmental Management – New Work Item Proposal – Material Flow Cost Accounting, ISO/TC 207/SC N 856, Geneva, December 2007.

International Standardization Organization, Environmental Management – Result of Voting on New Work Item Proposal – Material Flow Cost Accounting, ISO/TC 207/SC N 856R, March 2008.

Japanese Ministry of the Environment, Environmental Accounting Guidelines. Tokyo, 2002; http://www.env.go.jp/en/ssee/eag02.pdf.

Japanese Ministry of Economy, Trade, and Industry, METI, Environmental Management Accounting, Workbook. Tokyo, 2002. http://www.meti.go.jp/policy/eco_business/pdf/workbook.pdf

Japanese Ministry of Economy, Trade and Industry, METI, Guide for Material Flow Cost Accounting. Tokyo, 2007; http://www.meti.go.jp/policy/eco_business/pdf/EMA(MFCA)-english%20ppt.pdf.

Jasch Ch. Environmental Performance Indicators and Standard Framework of Accounts, How to Define System Boundaries and Reference Units, in The Green Bottom Line–Environmental Accounting for Management; Bennet M., James P., Hrsg. Greenleaf Publishing: Sheffield U.K., 1998.

Jasch Ch. Environmental Management Accounting, Procedures and Principles, United Nations Division for sustainable Development, Department of Economic and Social Affairs (United Nations publication, Sales No. 01.II.A.3) 2001; www.un.org/esa/sustdev/estema1.htm.

Jasch Ch. Environmental Management Accounting Metrics: Procudures and Principles, in Bennet M., Bauma J. and Wolters T. (Eds.), Environmental Management Accounting: Informational and Institutional Developments: Kluwer Academic Publ.: Dordrecht, NL, 2002.

Jasch Ch. Environmental Management Accounting (EMA) as the Next Step in the Evolution of Management Accounting. Journal of Cleaner Production, Volume 14 Number 14, 2006a.

Jasch Ch. How to Perform an Environmental Cost Assessment in One Day. Journal of Cleaner Production, Volume 14 Number 14, 2006b.

Jasch Ch. Funding Options for SMEs to Finance CP Projects and EST Investments, UNIDO, Vienna, 2007.

Jasch Ch., Danse M. Environmental Management Accounting Pilot Projects in Costa Rica, in M. Bennet, P. Rikhardson, S. Schaltegger (Eds.) Implementing Environmental Management Accounting: Status and Challenges. Kluwer Academic Publ.: Dordrecht, NL, 2005.

Jasch Ch., Lavicka A. Pilot Project on Sustainability Management Accounting with the Styrian Automobile Cluster. Journal of Cleaner Production, Volume 14 Number 14, 2006.

Jasch Ch. and R. Rauberger. A Guide to Corporate Environmental Indicators. On behalf of the German Federal Ministry for the Environment and the German Federal Environmental Agency in Bonn, December 1997.

Jasch, C. and H. Schnitzer. Umweltrechnungswesen – Wir, zeigen, wie sich Umweltschutz rechnet, Beispielsammlung zur Umweltkostenrechnung und Investitionsrechnung. Vienna: Bundesministerium für Verkehr, Innovation und Technik and Bundesministerium für Land- und Forstwirtschaft, Umwelt, und Wasser, 2002.

Jasch Ch., Schnitzer H. Environmental Management Accounting – Pilottesting and Case Studies, on behalf of the Ministry for Transport, Innovation and Technology as well as the Ministry for Agriculture and Environment, Published as Research Report 29/02 of the IÖW, the Methodology Part is Available for Download at www.ioew.at. 2002

Jasch Ch., H. Dimitroff-Regatschnig, and H. Schnitzer. Entwicklung eines methodischen Ansatzes zur Integration von Umweltkosten in das betriebliche Rechnungswesen; Hrsg. Bundesministerium für Umwelt, Jugend und Familie, Wien 1997; Schriftenreihe 24/1997 des IÖW Wien (available only in German).

Kneese A., R. Ayres, and R. d'Arge. Economics and the Environment, a Materials Balance Approach. Resources for the future Inc.: Washington, DC, 1970.

Kokubu K. and M. Nakajima. Sustainable Accounting Initiatives in Japan: Pilot Projects of Material Flow Cost Accounting in J. D. S. Hausmann, C. Liedtk and E. U. Weizsacker (Eds.) Eco-Efficiency and Beyond. Greenleaf Publishing, pp.100–112, 2004.

Kokubu, K. and E. Nashioka. Environmental Management Accounting Practices in Japan, in P.M. Rikhardsson, M. Bennett, J.J. Bouma, and S. Schaltegger (Eds.) Implementing Environmental Management Accounting: Status and Challenges, Springer, pp. 321–342, 2005.

Kyoto Protocol; http://europa.eu.int/comm/environment/climat/kyoto.htm, 1997.

LfU, Landesanstalt für Umweltschutz, Baden-Würthemberg (Hrsg.), Betriebliches Material- und Energieflußmanagement, ÖkoEffizienz durch nachhaltige Reorganisation, Karlsruhe, Oktober 1999 (available only in German).

METI, Ministry of Economy, Trade and Industry, Guide for Material Flow Cost Accounting Japan, 2007, www.jmac.co/jp/mfca

Montreal Protocol on Substances that Deplete the Ozone Layer, 16 September 1987; www.ozone.unep.org/Ratification_status/montreal_protocol.shtml.

Munkoe L., C. Jasch. Waste Reduction Program Based on IFAC's EMA Guideline in Danisco A/S, in S. Schaltegger, M. Bennet, R. Burrit, C. Jasch (Eds.) Environmental Management Accounting for Cleaner Production, Springer: New York, 2008.

Murauer Website; http://www.murauerbier.at/. Environmental Statement of Obermurtaler Brauerei 2003

Pojasek, R. Practical Pollution Prevention – Understanding a Process with Process Mapping, Pollution Prevention Review (Summer 1997a); http://www.pojasek-associates.com/Reprints/understanding-a-process-with-process-mapping.pdf.

SCA Laakirchen Environmental Statment; http://www.sca.at.

SEEA, System of Environmental-Economic Accounting, United Nations Statistical Division, European Commission, International Monetary Fund, Organization for Economic Cooperation and Development and World Bank, Handbook of National Accounting: Integrated Environmental and Economic Accounting, SEEA, 2003.

Schaltegger, S. and R. Burritt. Contemporary Environmental Accounting: Issues, Concepts and Practices. Sheffield, UK: Greenleaf Publishing, 2000; http://www.greenleaf-publishing.com/pdfs/ceach1.pdf.

Schaltegger, S., K. Müller and H. Hinrichsen. Corporate Environmental Accounting. Chichester, UK: Wiley, 1996.

Sprenger R., Die amtliche Umweltstatistik in der Sackgasse?, Ökologisches wirtschaften 1.2007, Institut für ökologische wirtschaftsforschung, Berlin, 2007 (available only in German).

Staniskis J., Z. Stasiskiene, and C. Jasch. Assessment of Environmental Costs for Sustainable Industrial Development, Monograph. Kaunas Technologija, Kaunas, 2005.

Statistisches Bundesamt: Statistisches Jahrbuch 2006 für die Bundesrepublik Deutschland, Wiesbaden, 2006 (available only in German).

Strobel, M., Systemisches Flussmanagement. Flussorientierte Kommunikation als Perspektive für eine ökologische und ökonomische Unternehmensentwicklung (Dissertationsschrift), Universität Augsburg 2000 (available only in German).

Strobel, M. and C. Redmann. Flow Cost Accounting, IMU (Institute für Management und Umwelt), Augsburg 2001.

United Nations Statistical Division, European Commission, International Monetary Fund, Organization for Economic Co-operation and Development and World Bank, System of National Accounts, 1993; http://unstats.un.org/unsd/sna1993/introduction.asp.

United Nations Conference on Trade and Development. Accounting and Financial Reporting for Environmental Costs and Liabilities (UNCTAD/ITE/EDS/4). New York and Geneva: United Nations Publications, 1999; http://www.unctad.org/Templates/webflyer.asp?docid = 205&intItemID = 1397&lang = 1.

United Nations Conference on Trade and Development. A Manual for the Preparers and Users of Eco-Efficiency Indicators (UNCTAD/ITE/IPC/2003/7). New York and Geneva: United Nations Publications, 2004; http://www.unctad.org/Templates/webflyer.asp?docid = 4371&intItemID = 1397&lang = 1.

United Nations Division for Sustainable Development. Environmental Management Accounting, Procedures and Principles, New York and Geneva: United Nations Publications, 2001; http://www.un.org/esa/sustdev/sdissues/technology/estema1.htm.

United Nations Division for Sustainable Development. Environmental Management Accounting: Policies and Linkages. New York and Geneva: United Nations Publications, 2002; http://www.un.org/esa/sustdev/sdissues/technology/estema1.htm.

United Nations Environment Program and United Nations Industrial Development Organization. Audit and Reduction Manual for Industrial Emissions and Waste, Paris, 1991.

United Nations Statistical Division, European Commission, International Monetary Fund, Organization for Economic Co-operation and Development and World Bank, System of National Accounts, 1993; http://unstats.un.org/unsd/sna1993/introduction.asp

United Nations Statistical Division, European Commission, International Monetary Fund, Organization for Economic Co-operation and Development and World Bank, Handbook of National Accounting: Integrated Environmental and Economic Accounting, 2003.

Verbund Sustainability Report 2003; http://www.verbund.at/en/group/sustainability/nachhaltigkeitsberichteng_2003.pdf.

Wagner, B. and S. Enzler (Eds.) Material Flow Management: Improving Cost Efficiency and Environmental Performance, Phsica-Verlag, Heidelberg, New York, 2006.

World Business Council for Sustainable Development. Measuring Eco-Efficiency: A Guide to Reporting Company Performance, Genf, 2000.

Index